SOU'WEST IN WANDERER IV

*Frontispiece*. On a broad reach with all sail set including the mizzen staysail, *Wanderer IV* can carry a total area of 1,640 square feet.

# Sou'west in Wanderer IV

ERIC C. HISCOCK

*Author of*
*Cruising Under Sail*
*Voyaging Under Sail*
*Beyond the West Horizon*
*etc*

WITH 58 COLOUR PHOTOGRAPHS
BY THE AUTHOR AND HIS WIFE
AND 8 CHARTS

LONDON
OXFORD UNIVERSITY PRESS
NEW YORK TORONTO
1973

*Oxford University Press, Ely House, London W.1*
GLASGOW NEW YORK TORONTO MELBOURNE WELLINGTON
CAPE TOWN IBADAN NAIROBI DAR ES SALAAM LUSAKA ADDIS ABABA
DELHI BOMBAY CALCUTTA MADRAS KARACHI LAHORE DACCA
KUALA LUMPUR SINGAPORE HONG KONG TOKYO

ISBN 0 19 217528 9



*Printed in Great Britain*
*by W & J Mackay Limited, Chatham*

For
SUSAN

# CONTENTS

# ILLUSTRATIONS

## CHARTS



## PLANS

# ACKNOWLEDGEMENT

The acknowledgements of the author are due to the Editors of *Sail* and *Yachting World*, in which magazines some parts of this story were first published.

# 1

# A Home Afloat

Night had fallen, and the lights of Holland sparkled in strings and clusters beneath us as the pilot of our plane started the let-down, and we, obeying the illuminated sign, fastened our seat-belts. The halls of Amsterdam's airport engulfed us. On a moving walkway, almost a magic carpet, we glided to the blue-and-white-uniformed immigration officials, who glanced briefly at our passports, and the customs, who nonchalantly chalked our bags and let us go. The place was strangely silent and almost empty—our fellow passengers from London had vanished, as is the habit of people who are familiar with airports and Hertz and Avis—and we were the only passengers in the airport bus as it sped along the broad motorway and into the tram-tangled city, where on either side we could look through uncurtained windows into Dutch interiors, each one softly lighted by shaded lamps. In contrast the railway station was bursting with light and life, as was the diesel train where garrulous students packed us in while we roared and rattled through the night.

'Hoorn', said a guttural voice in a loudspeaker at the end of the coach, and as the train drew in we descended to the low platform of an almost blacked-out station, and groped our way to an equally dark bus, which after the driver had switched off the few dim lights and tried the starter several times, burst into vibrating life and bumped us through the dark countryside. Miles from any place, it seemed, we were told to get out, and for a while stood shivering in the chill winter wind at the roadside until another bus drew up, embarked us and a few cigar-smoking Dutchmen and their round jolly-looking women, and we lurched on once more past windswept dyke and field to

Medemblik, a small, industrial town near the northern end of the Ijsselmeer. A lean hotel near the bus stop accepted us with lifted eyebrows, for this was probably a rep's retreat unaccustomed to taking in women. Our bedroom with its sloping, highly polished floor and twin basins smelling of drains, struck chill after the warmth of the bar below, where we had stood ourselves a drink and watched the billiard players. Our cosy home at Yarmouth on the Isle of Wight, which we had left that morning, now seemed a very long way away.

After an early breakfast we walked along the frosty, cobbled street beside the canal where buxom fishing vessels lay, to the shipyard office. The Builder was not yet in, but after we had explained who we were the clerk said we could go into the shop and look round. This was a vast, echoing place containing several partly completed vessels and was heated by giant iron stoves, and there for the first time we saw the massive hull of the ship which was soon to be our floating cruising home—a swelling erection of grey and rust-brown steel plates joined one to another by lines of welding. Instead of the pleasant scent of sawdust and planings and the music of saw and caulking mallet, to which we had grown accustomed during the building of our two earlier wooden yachts, the cold, bitter smell of steel stung our nostrils, and the harsh rasp of metal on metal and the crackling hiss of welding torches bombarded our ears.

'How do you like her?'

I turned to confront the Builder, a big, fresh-faced man probably in his forties, with an infectious grin.

The scream of a grinder drowned my reply.

'She's beeg, this ship of yours. How many crew men will you have?'

I said there would be only my wife and myself. He looked at me as though he thought I was crazy.

I glanced again at the huge steel erection surrounded by scaffolding, which towered above us with its full-bodied sections and out-thrust clipper bow—and I began to wonder. Perhaps I *was* crazy to suppose that the two of us could manage with competence and safety and maintain in reasonable condition anything so large. Then I looked at Susan and was relieved to see that she was actually smiling.

We returned to the small, superheated little office, and having shaken hands all round, sat down with the Builder and his clerk and the Designer and his draughtsman, who had just arrived from a distant town, to discuss the many problems that had already arisen over the designing and building of our dreamship.

For several years Susan and I had been talking about selling our home and its contents and our beloved *Wanderer III* and of getting a vessel large enough to be our permanent cruising home, for after twenty years of house ownership we had come to the conclusion that boats did not mix too well with houses and gardens; both called for constant attention and neither got it. Probably we would have continued to discuss this exciting idea until we grew too old to do anything about it, had it not been for two letters we received in the spring of 1967 while we were cruising on the east coast of the U.S.A. One informed us that our tenants had not paid any rent for the house for some considerable time, and were not likely to do so in the foreseeable future; the other, from the County Council, stated that a strip of our property was required for road widening. These, you may well consider, are small matters, but they served to topple us over the brink of indecision. With great enthusiasm, and the surprising but welcome encouragement of our accountant and bank manager, we picked over the brokers' advertisements, and wrote letters to designers and builders, determined that our floating home must be ready for us to move aboard and sail away the following year, and that everything must be set in train by the time we concluded our American cruise.

But quickly two things became clear: none of the second-hand yachts of which we had particulars appealed to us sufficiently, and to have one of the size we needed (about 20 tons displacement) designed and built in Britain would cost too much and take too long. Today, with so many excellent makes of car available, surely nobody would seriously consider having a special one designed and built, and presumably the same should apply to yachts; but we could not find a British stock design (for construction in wood, steel, or glass-reinforced plastics) of the size and type we needed, and drawing not more than 5 feet 9 inches—such comparatively shoal draught was required so that our new yacht should be able to use any of the anchorages that *Wanderer III* had enjoyed inside the one-fathom line. Also, oh vanity! we wanted one looking just a little more exciting and characterful than the usual run of modern designs, and to our eyes this meant, among other things, a lively sheer and a clipper bow with carved trailboards.

Then an advertisement in an American magazine caught my eye and held it. This was for a stock design by a Dutch architect, and seemed to meet many of our requirements, including moderate draught and clipper bow, and after we had seen a print of the lines we felt that this was the ship for us, or as near to our ideal as we were likely to get without having a special design drawn.[1] Naturally there were some things we would have liked different: the waterlines seemed a bit full aft, four tons of ballast appeared to be meagre for so large a vessel, and I particularly disliked the partly balanced rudder hung abaft an enormous propeller aperture; neither would we have chosen ketch rig or a central cockpit. However, we learnt that if we left these things as they were and had the yacht built of steel in Holland, the cost would be what in our more optimistic moments we thought we could just afford after we had sold our house and other belongings, and this would be about one-third less than if the yacht were built of wood in England. So long as we did not alter the position of the bulkheads we could arrange the accommodation details as we wished. We would have to keep the list of 'extras' down to the absolute minimum, so we tried to forget about a beautiful teak deck laid over the steel, for this would have cost an additional £1,000.

I made a lot of sketches and exchanged many long letters with the consultants in London, a firm which would make some mysterious, but apparently legal, arrangements for our sterling to be converted into guilders, and at the same time cover us by forward buying against the risk of devaluation of the pound. A shipbuilder at Medemblik on the Ijsselmeer agreed to build the yacht and said he would have her ready in time for exhibition at the Amsterdam boat show in March. After each of her world voyages *Wanderer III* had been exhibited at the London boat show, and Susan and I are of the opinion that boats are best left

[1] Details and plans of *Wanderer IV* will be found in the Appendix.

in their proper element; the journey by road, the handling by crane, and the wear and tear of many visitors' feet are not good. But, as the Designer pointed out, if we allowed the Builder to exhibit her we could at least be certain of one thing: he would have her ready in time.

On our way back from the U.S.A. to England we called at the Azores and there signed the contract, feeling a little sad that this could not be with our old friends William King or A. H. Moody. You might suppose, as we did, that everything was set fair, and that provided we paid the instalments as they became due, building would progress smoothly. But it did not turn out quite like that.

At an early stage the Designer asked how much water and fuel we wanted to carry; I said it would be fine if we could have between 200 and 300 gallons of each, but that as we were buying a stock design we would of course accept the tankage provided, and asked how much that was. He did not reply, and unknown to us at the time he re-drew the lines so as to produce a hull of greater displacement capable, he hoped, of accommodating our needs. He did this with the best will in the world and at no extra cost, but somehow the sole in the fore half of the accommodation got raised, and with it the coachroof; we did not like the look of this, and our ideas for accommodation details, which I had drawn on the original set of plans, now did not fit properly.

So worried and depressed did Susan and I become (this seems to be a common condition among those having yachts built, and attacks me most strongly at 3 o'clock in the morning) that we twice tried to cancel the contract; but neither Builder, Designer nor the consultants would let us slip, and we are glad now that they did not, for in time most of the difficulties were settled to our satisfaction.

But new problems arose. The Builder's foreman joiner had a heart attack, and a few weeks later his clerk, the only man in the place with a good understanding of English, was involved in a car smash and got concussion. I was therefore asked to order some of the British items of equipment, and it was a pleasure to have dealings with old and trusted firms such as Cranfield Sails, Blakes of Gosport, Henry Browne, and Taylors Para-Fin. But I found that some other firms were apparently not interested in selling anything; letters remained unanswered, promises made when I telephoned were not kept; so the weeks ran into months, Christmas came to cause further delay, and then the London boat show. We had some entertainment over the business of getting navigation lamps. The only kind available in Holland had dioptric lenses, and these are unsuitable for sailing craft because when heeled the narrow plane of light they produce shines up into the sky or down into the water; we wanted lamps with plain glass, and ordered a set—port, starboard, and stern—from an English firm which specialized in this work. I think my order must have got into the clutches of a computer, for I received many pieces of paper but no lamps in spite of repeated urgent requests. But eventually I succeeded in getting in touch with the managing director; he was apologetic and explained that the delay was caused by the fact that as *Wanderer IV* (the Registrar of British Shipping had by now approved this name) was to be exhibited at the Amsterdam boat show, they wanted the lamps to be as perfect as possible; unfortunately the chrome-plating on one was not quite up to the usual high standard and had to be done again. This surprised me, for I had not ordered chromed lamps, and when at long last they arrived we found they were indeed not chromed but painted. We also discovered that the stern lamp shone over much too wide an arc and was in fact a steaming lamp. The glass of the starboard lamp had been broken, either in transit or at the yard. Nobody said a word, but someone fitted in its place a piece of talc and painted it with dark green paint through which no ray of light could penetrate. I only discovered this just before we left on the passage to England, and therefore had to borrow a Dutch lamp. This of course had a dioptric lens, and as the bulb holder had been fitted incorrectly well above the centre of it, the only light that lamp shed was a small, green pool on the water immediately beneath it. I have mentioned this little business at some length just to illustrate the kind of problem, and there are many such, that has to be dealt with in the building and equipping of a small vessel today.

When I was not too occupied with such things as this I gave some thought to the bigger issues of design and build and hoped that our decision to have this vessel built was the right one. By today's standards her keel does look rather long and straight—there is about 20 feet of it—but at least I thought it should make slipping or taking the ground less of an anxiety, and in lateral resistance should go far in making up for lack of draught; a possible dignity of movement when going about should not concern us as we did not have any form of racing in mind.

The clipper bow is not only a vanity; I regard it as a good, dry end to a vessel of this size, and together with the bowsprit, it lengthens the base of the sail-plan so that sufficient area of sail may be had without indulging in tall masts. For ocean sailing I feel this is important, as the strains on the rigging are less, and low sails do not get thrown about by the motion quite so much as high ones; also the mainmast is short enough to pass under the fixed bridges of the Intracoastal Waterway of America should we ever decide to go there again.

I was not sure how we would like the knuckle where topsides and bulwarks meet at the bows, for in the traditional type of clipper bow the flare is continuous; but I was sure the Designer was right to introduce the knuckle in this instance, otherwise the flare of the bulwarks would have been excessive, like that of a Spanish sardiner. As the bulwarks were to be painted dark green and the topsides white, I thought the knuckle might not be very noticeable. I would have liked a little more sheer, and if only I had known that the lines were to be re-drawn I might have asked for this together with the other modifications I have already mentioned, though I do feel there are limits to what one should suggest the artist does with his masterpiece.

If machinery and electrics are required, and we wanted both for our comfort and convenience, it seems unreasonable to place them aft under a stern cockpit or doghouse, where they will probably get wet, their weight be in the wrong place, and access to them difficult. We therefore reconciled ourselves to the centre cockpit, though we had misgivings about its size, which was dictated by the width of the superstructure and the position of the steel bulkheads enclosing the engine-room below, for in the event of a sea falling into it it would hold a great weight of water (about 4 tons) some of which would surely find its way down the companionways into the accommodation. We would have preferred to dispense with the companionway into the sleeping cabin, for it would be vulnerable to spray and rain, and have instead a passage through beside the engine from the forward to the aft accommodation, using headroom provided by one of the fore-and-aft cockpit seats. But this was not on the plans, and certainly the completely isolated engine-room would ensure that heat, smell and noise were kept away from the rest of the ship.

Oak, iroko and copper are the building materials which Susan and I felt we knew something about, and it was not easy to adjust to the thought of steel construction. Our chief worry in this connexion was about the risk of galvanic action between dissimilar metals, for all the sea-cocks are of yellow metal, and this worry increased when we discovered that instead of providing cone-type cocks right up against the plating, as was called for in the specification, gate-valves had been fitted on stand pipes which varied in length from a few inches to several feet. The point of having sea-cocks was thus defeated, for a fractured or corroded pipe between the plating and a cock could sink the ship. However, we had sacrificial zincs fitted in positions recommended by a firm that specialized in metal protection, and hoped for the best. Rusting ought not to occur if the preparation (sand- or shot-blasting) and painting with the right materials in the right weather conditions are honestly done, but there are a few places which are notoriously difficult in this respect, notably under the bulwark capping, and in the waterways; but at his own suggestion and expense the Builder made the flat top of the bulwarks, on which the teak capping was to be fastened, of stainless steel, and at my request dispensed with the usual waterways and carried the teak deck—he made us a present of this because, I think, he just could not bear to exhibit the yacht in Amsterdam without it—right out to the bulwarks.

Condensation down below would not be the bother it is in some steel vessels, for the underside of the deck and the inside of the skin plating in way of all the accommodation was to be ▶

1. Displacing all of 20 tons, she is a big vessel for the two of us to maintain in good order. We have found the clipper bow to be more than just a vanity.

insulated with glass wool entirely boxed in with plywood; but no doubt I would fuss (at 3 a.m.) about the places I could not see or reach where water might collect and electricians might have left their snippets of copper wire, for this creates an action detrimental to steel.

Steel, of course, has certain merits, notable among them being great strength and elasticity, no leaks or attack by marine borers; more space (this is particularly noticeable with tanks); cost is lower than with wood construction, and a one-off can be built comparatively inexpensively, which is not so with glass-reinforced plastics construction. I hoped we might discover other advantages in due course.

I have never been an admirer of the ketch rig, believing it to be inefficient, and that because on some points of sailing the mizzen interferes with the mainsail, or the mainsail with the mizzen, one or other has to come in, and then the vessel is under-canvased and probably out of balance. However, the design was for a ketch, and Susan and I were prepared to try it for we thought, mistakenly as we soon discovered, that the split-up rig and therefore smaller individual sails would be easier for us to handle than those of the cutter rig. But instead of carrying the one big headsail which was shown on the original plan, we had a double headsail rig of jib and staysail (Plate 2). This called for two pairs of crosstrees instead of one, and a pair of running backstays, which are always a bother, to take the pull of the staysail and keep its luff taut. I decided, rather against my first inclination, to have the staysail on a boom for ease of handling when tacking. As the mainsail would be only 350 square feet in area and its boom provided with roller-reefing gear, we felt we should be able to handle it in all conditions with fair ease. In addition to the working sails we ordered a 500-square-foot light weather headsail, a mizzen-staysail of 400 square feet, and a pair of twin running sails.

Having on our final cruise in little *Wanderer* enjoyed one of Blondie Hasler's excellent vane steering gears, we hoped never

◀

2. We had never been admirers of the ketch, but *Wanderer* was designed with that rig and we were prepared to try it. Here she is moving fast under all plain sail including No. 2 jib.

to have to steer by hand for long periods again. But it seemed that our new ship would be a bit too big for one of Blondie's standard gears, and something special would have to be built at considerable expense and with no real guarantee of success; also the mizzen would have to be reduced in size to make room for the vane to swing. I hoped that with the split-up sail plan it might often be possible to achieve self-steering, but for the occasions when that was not possible we decided, with some misgiving, to install an electric automatic helmsman.

The engine specified was a 93 h.p. (continuous rating) marinized Ford diesel, but I felt that this was excessive power for a sailing vessel displacing 20 tons; we finally settled for the 61 h.p. Ford, and thereby saved almost enough money to pay for the installation of a 50-amp electricity generating plant with which to charge the batteries.

Below deck the fore part was intended for the stowage of sails and anchor chain, and for use as a workshop/darkroom. The saloon (Plate 3, *top*) calls for little comment except that the forward bulkhead was to be provided with bookshelves arranged round a recess for our David Cobb painting of *Wanderer III*. Heating was to be by a Kempsafe diesel stove, and we hoped that the chart table aft on the starboard side would have sufficient stowage under it for our quite considerable collection of charts. The floor space in the galley was intentionally kept small to prevent the cook from being flung about too much in rough weather (Plate 3, *bottom*). Normally it would be provided with a gas cooker and an electric refrigerator, but because of the difficulty of obtaining gas in some remote (and some not so remote) places, we decided to do as we always have done and cook on a Taylors Para-Fin stove swung on its fore-and-aft axis. We knew, after sharing anchorages with yachts of the West Indies charter fleet, that electric- or mechanically-operated refrigerators call for much engine running, often several hours a day; so as one or two of our cruising friends, including the Caldwells in *Outward Bound* and the Guzzwells in *Treasure*, had successfully used paraffin-operated refrigerators when in port, we decided to try one of these. Electrolux appeared to be the only firm in Britain to make them, but to our astonishment refused to sell us one for use afloat because of the fire risk—this seemed to us just as unreasonable as a maker of navigation instruments refusing to sell a compass to someone who drinks beer in case a magnetic beer can should be left standing near it. However, the builder managed to get hold of one and put it in.

Aft of the cockpit is what agents call the 'owner's suite' comprising a two-berth sleeping cabin (Plate 4, *left*) with lockers, dressing table, and an enormous wardrobe, and adjacent to it a heads and shower compartment (Plate 4, *right*) all gussied up with teak gratings, stainless tray with electric pump, fan, basin, mirror and soiled linen locker [*sic*]; but we began to wonder if perhaps the shower part of this was going to be a white elephant except in hot weather. According to the 10-page specification and the plans, there should be installed in the engine-room a calorifier through which cooling water from the engine passes to heat water for the shower. Clearly it is a disadvantage to be able to have a hot shower only, say, on Friday evening after charging the batteries, but that is better than no hot shower at all when living permanently on board. But the Designer suddenly informed me that calorifiers were not satisfactory, and had been known to cause damage to the engines with which they were connected, and rightly refused to put one in. The only alternative was to have a gas heater installed, which would mean carrying a third type of fuel for that single purpose, or else use a saucepan, a bucket and a sponge.

Problems of one sort or another continued to arise, some of them due to language difficulties, and it was unfortunate that we could not afford the time, even if we could have obtained the money in a period when one was allowed only £50 a year of foreign currency, to go and stay at Medemblik, watch over the building and supervise the many little details which are so important but cost no more to do right than wrong; also we would have liked to get to know the craftsmen who were fashioning our new home. In fact we paid only one more visit, when in a blizzard we took over a car-load of the things we would need for the trip across the North Sea to England.

It was towards the end of April that we travelled by night ferry to the Hook, and were met there by David Guthrie who was at that time living in Holland. We had last seen him in Antigua, where with his corgi Cider he was making a circuit of the North Atlantic in his sloop *Widgee*. He kindly drove us to Medemblik, where we found our new ship lying afloat at her builder's yard looking immaculate and, we thought, very handsome. She had but recently returned from her *tour de force* at Amsterdam where, we were told, she had been proclaimed the yacht of the year and had been greatly admired by many visitors. As we intended to install ourselves aboard immediately, Susan went shopping for food; I, meanwhile, strolled round the spacious decks, admiring the curved teak planks and the superb finish everywhere; then casting my eye rather critically aloft, I noticed that most of the running rigging, and some of the standing rigging, had been incorrectly rove. However, I was not dismayed or even surprised at this, for as Conor O'Brien pointed out many years ago, no yard rigger is ever in much danger of having to go to sea in the vessels that he rigs.

I lifted the forward, starboard seat in the cockpit, switched on the lights and made my way down the steel ladder into the engine-room (Plate 5). This 8-feet-by-12-feet compartment was all decked out in white enamel against which the metallic-green machinery stood out boldly. With its network of pipes and scarlet cocks it looked impressive, and it was clear that I would need some instruction before I interfered with anything down there. I climbed out, closed the hatch and took a look at the accommodation. Wonderfully spacious it seemed after little *Wanderer*'s, as I opened and closed the many well-fitting locker doors and ran my fingers along the polished bookshelves. Perhaps I was over-critical that day, but I did notice that most of the fiddles, though beautifully finished, had been made with their inner edges sloping as though specially designed to throw their contents as quickly as possible onto the deck, and I thought that too much of the finish was in polished teak and not enough in white enamel, for in spite of the perspex hatches and many large oblong ports, it was a little gloomy down below. But the general impression was of space and comfort and there was nothing there that we could not alter to meet our own fancy, given time.

Susan speaks no word of Dutch, but when she returned she brought with her all we would need in the way of food and drink for some days; she soon had the cooker going, and we sat down to our first meal at the saloon table, which is so large that we subsequently found we could seat eight people at it.

It was strange and rather exciting to realise that we would shortly own a ship so large and glamorous, but we did not waste too much time daydreaming, for Susan had plenty of housewife's work to do below, and I needed to sort out the rigging to the best of my ability before we sailed on acceptance trials in two days' time. Builder's trials had already been sailed (with the jib set upside down, as we could see in the photographs) and I discovered that several of the terylene halyards had been damaged as though by heat where, incorrectly rove, they had been forced across the sharp strips of brass with which the edges of the masthead sheave-holes had surprisingly been shod. These had to be blunted with a file, and subsequently we removed them. But more serious than this was the fact that the angle of the tangs beneath each pair of crosstrees was incorrect, so that the shrouds did not lead from them in a straight line. Below Susan found that what was to have been her best galley locker, for which we had specified shelves of the correct height to accommodate her pots and other gear, had been filled up with the main switchboard, and that the locker under the sink, which was to have held the refuse bucket, had apparently been built solely for the convenience of the plumber, who had run the waste pipe right through the middle of it. And when it rained, which it often did, most of the 13 opening ports leaked, and the drip-catchers beneath them, because they had not been bedded, let the water straight through on to the bunks and settees.

Two days later the Designer and the Builder joined us for trials; the Ford was started, and running sweetly took us out into the almost windless Ijsselmeer. I thought it a little odd that although the propeller is left-handed the ship wished all the time to turn to port. On asking about this I was told that the propeller is right-handed, but it is not. There is a brake on the shaft to hold it from rotating when under sail, but as there was so little wind we had not bothered to engage it. While in the engine-room I noticed that each time the shaft rotated a squeak came from its flexible coupling, which suggested that the engine was not in line with the shaft. The experts did not take this at all seriously, and declined to do anything about it. The trials under sail were disappointing through lack of wind, and we could form no opinion of the performance, though for what it was worth it did seem as though the ship wanted even then to turn to port. It was also disconcerting to find that the auxiliary generating plant firmly switched itself off after running for a short time, and refused to supply any more current until its control box had cooled off. The Builder, with his usual delightful smile, shrugged this off with the remark:

'It just mean the batteries are fully charged.'

I did not believe him, and neither, I think, did the Designer, for whom I was feeling a little sorry as he looked so worried that day.

We returned to the yard, where a half-hearted attempt was made to cure some of the smaller defects, but it soon became clear that nothing much was going to be done about the major ones; so instead of remaining there and insisting, as I suppose we ought to have done, we paid the final instalment, all 64 of the ship's shares became ours jointly, and on 3 May the Designer took the wheel, and for 5 hours we motored south across the shiny, windless Ijsselmeer, which year by year is growing smaller as areas of it are reclaimed, to spend the night in the charming little harbour of Muiden, which very soon will no longer exist. The Designer kindly gave us a fine dinner ashore, and the next day we motored on for 35 miles in heavy rain through the North Sea Canal, where we did not seem to handle too well in the lock, and tied up to the quay at Ijmuiden at the canal's western end. The Designer and his charming wife, who had come to collect him, wished us good luck and said good-bye.

For several days of wind and rain we remained weather-bound, unwilling to take our untried vessel out into the North Sea until more gentle conditions prevailed, and we were glad of the comforting warmth from our diesel heater, and I was not sorry to have this time to find out more about the array of mechanical and electrical equipment. Among all this lot there were at least two familiar items, the Blake heads installed aft and in the fore part of the ship. So when one of these went out of action, I knew how to strip it down; I found a rusting 2-inch by $\frac{3}{8}$-inch steel bolt lodged on the seating of the lower valve and further investigation disclosed its twin lurking in the soil pipe ready to take its place. Surely it must have been a very deaf plumber who dropped these into the china bowl without knowing about it. We still could not charge the batteries with the auxiliary plant for more than a few minutes at a time, but as we had run the main engine, which also charges, so much recently this was of no real importance yet.

As soon as the weather looked reasonable we put to sea, and it was unfortunate for our fenders and our lines that the edges and the bollards of the sea-lock had that morning been painted black.

Our intention was to sail direct to Yarmouth in the Isle of Wight, 250 miles distant, because arrangements had been made with customs there for the temporary importation of the yacht so that we would not have to pay the required 10 per cent duty on her, provided we left the country within a year. (Sometimes I wonder if we shall become a sort of Flying Dutchman.) We planned to keep outside the banks, and sail from light-vessel to light-vessel. The first of these, the Goeree, is about 40 miles from Ijmuiden, and we headed for it in a confused short sea with a light beam wind, but after we had run our distance by our old and trusted Walker's log (this we had used for about 150,000 miles in earlier yachts) there was no sign of the light-vessel. We had with us a Heron/Homer direction-finding radio set, which we had had in *Wanderer III*, but to use this in the normal manner in a steel vessel is impossible, not just because of the unknown deviation of the hand-held compass, but because of the closed loop effect on the aerial. The only way in which it can be used in a steel vessel is to line up its aerial on the fore-and-aft or athwartships line, and get the null by swinging the ship so that the beacon lies ahead, astern, or on either beam, and then read the steering compass which, presumably, has been adjusted for errors caused by the steel surrounding it. However, using this method my round of bearings made absolute nonsense, and we were beginning to feel completely lost when Susan spied the silhouette of the training schooner *Sir Winston Churchill* on the horizon to leeward of us. The schooner was close-hauled and going fast, although there was not a lot of wind, and only by running our engine to help the sails were we able to close with her. As we crossed her wake I asked for the bearing and distance of the Noord Hinder (the next light-vessel along our route), and a voice replied:

'If you wish to speak with me, come under my lee.'

I wondered just how we were to do that, for the schooner was sailing so much faster than ourselves; but her master, Commander Willerby, luffed a little to allow us to come within comfortable hailing distance. A magnificent sight his vessel made in the grey of that overcast evening, her towering sails almost overshadowing us, a bone in her teeth, and the crew's life-jackets making splashes of bright orange along the bold, black sheer of her bulwarks. Willerby kindly gave us the information we sought, and we went our respective ways, the big schooner beating up for Rotterdam, our ketch bearing away for the Noord Hinder. The bearing and distance of this light-vessel, as provided by *Sir Winston Churchill*, showed that we were 15 miles off course after a run of only 40—a compass error of about 20°. *Wanderer IV* had been swung before leaving her builders, and the compass adjuster had complained of the difficulty of doing his job due, he said, to the magnetic parts in the bevel box of the Mathway steering gear close under the compass. Later, when we discovered, after someone had given me a magnet, that the boss of the steering wheel was of iron instead of brass, we thought this the more likely cause of the trouble, and replaced the wheel with one having a brass boss; but the Dutch adjuster was right, to some extent, though we did not discover this until we checked the compass ourselves in California after an English and an American adjuster had had a go at it, and found that half a turn of the wheel moved the card several degrees.

Now that we knew the approximate deviation on our course we were able to find the Noord Hinder. There was little wind by then, and we had the engine running continuously. We did not sight the Sandettie, for visibility was down to a mile or so in

▶

3. *Top*: The saloon gives us plenty of room in which to live and work and entertain our friends. The diesel heater with its chimney can be seen in the left foreground, and on the right a corner of the chart table, which has stowage under it for our collection of 800 charts. *Bottom*: Although its floor space is small so that the cook cannot get thrown about in bad weather, the galley has a large working area and plenty of lockers. The refrigerator (its vent can be seen close to Susan's back), and the swinging cooker both burn paraffin. The sink is amidships.

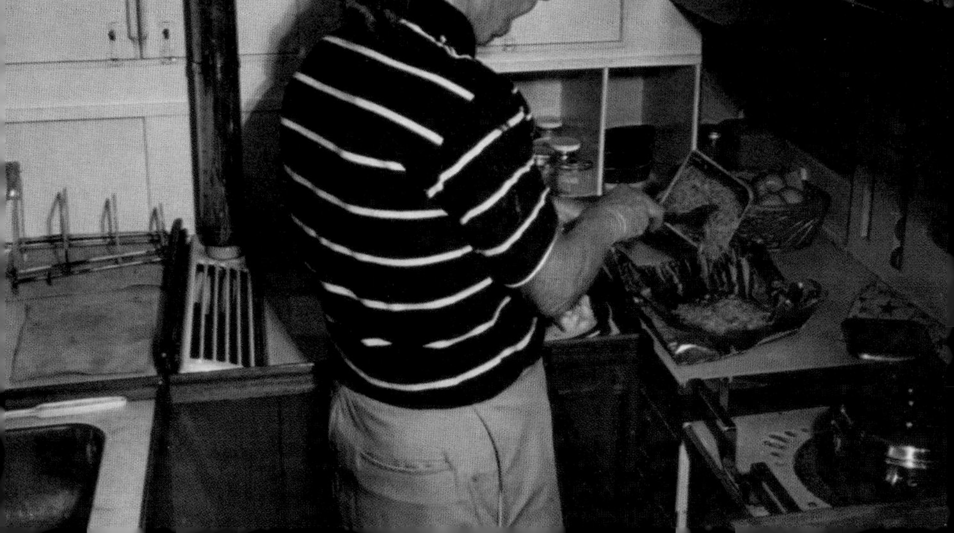

rain, but a succession of tankers and other commercial craft looming up ahead or astern showed us the way to the Strait of Dover, the busiest stretch of water in the world, where fortunately the weather cleared.

The engine is fitted with Borg Warner Velvet Drive transmission, a beautiful piece of hydraulic machinery incorporating 2-to-1 reduction gear, and I was dismayed to find that an oil leak had developed at the joint between the ahead and reduction housings; every few hours I had to pour in more oil. Then trouble with the fuel arose, so that several times each hour the engine faltered and lost power, though it never stopped. This turned out to be dirt from the fuel tank, stirred up by the motion, getting into and partly choking the supply pipe before this reached the filter. There were one or two other little problems as well. Both heads, because their soil pipes had not been carried up sufficiently above the waterline, tended to flood, and we realized that when going to sea we must always remember to close their sea-cocks as well as the cock on the wash-basin drain. Apparently the fresh water pressure system contained a lot of dirt; the motion stirred this up, too, so that the water coming from the taps was thick, like cement-wash, which in fact it was. Also, as the pressure tank was mounted athwartships instead of fore-and-aft, and its outlet was at the starboard end, when the ship was heeled to port, air instead of water came out of the taps.

We plodded thoughtfully on down Channel, and at the Owers met a suddenly arising force-7 westerly wind against which we tried to beat to reach the Solent. But the halyards had stretched, and the rain had so swollen the unprotected sheave holes in the mastheads that the sheaves jammed and it was impossible to get the sails properly set; also the main boom split from end to end along the glue lines, and only the roller-reefing gear fittings held it together. So we motor-sailed, and with the engine becoming increasingly dot and carry we managed to

◀

4. *Left*: The starboard side of the sleeping cabin. As the built-in bunkboards are too low, an extra one can be drawn out when at sea; one of these, painted white, can be seen alongside the dressing table. *Right*: The after heads compartment, with Baby Blake, wash basin, and shower facilities.

creep into the shelter of the Beaulieu River, where Pat Russell of Marine Services generously gave up his Saturday afternoon to making us sufficiently fit mechanically to motor on to Yarmouth when the wind eased.

I suppose that most new yachts have their little troubles, but *Wanderer IV* seemed to be having more than her fair share, and some of them were not so little. After customs had dealt with us and we had collected our clothes, books, charts and other belongings from our house, we got Harold Hayles to put in hand the making of a new boom, and then went to Moody's on the Hamble to see what they could do about some of our other troubles.

As the berthing master at Moody's marina took our lines he enquired after our cat Nicholson about whom he had read, and later searched the entire place and all the yachts in it when Nicholson worried us by overstaying his shore leave. As I had been answering Nicholson's fan mail, some of which protested that it was not fair of us to keep him confined afloat, I had better explain the situation. Since returning from our last voyage he had had the run of our garden and that of our neighbours as well as the adjacent farm; he proved to be a great hunter, and was popular because he helped to keep the mice, mole, and rabbit population down. Clearly he loved the life, so we arranged with our neighbours that they should look after him while we were away bringing the new ship from Holland, with the hope that he would take to them and become their cat. For three weeks he ate their food and slept on their beds, but never showed much appreciation, and when we got back to Yarmouth and Susan went to square up the house before the new owners moved in, his welcome was touching, he followed her about all day, and when she drove away each evening he sat dejectedly by the gate. Obviously he was her cat, so we brought him and his bag of wood planings aboard; he made much of us both, purred for a week, and settled down contentedly in his new home.

Each of the Moodys came to visit us in turn, and the facilities of their fine yard were immediately put at our disposal, and we found that the same friendly attitude among the men working there (I understand that none of them were union members) prevailed just as it did those many years ago when Susan's father's *Memory* and my first *Wanderer* laid up there. We had been told that the yard was very expensive, but we did not find that to be so.

Both masts were lifted out and their sheave holes enlarged and copper lined to keep the wet out, and Susan and I took the opportunity to give the masts a much needed coat of paint. New headsail sheet leads were made to replace the soft brass ones which had distorted under strain, and 12 defective stainless steel rigging screws were replaced with properly made ones of bronze. The suspected misalignment of the engine proved to be a fact, and this it was that had cracked the joint in the gearbox, which now had to go to its makers for repairs, and caused the oil leak. To get the alignment right the engine had to be lowered half an inch, and as all the available adjustment had already been taken up in Holland, four new feet had to be made for it. We went up on the slip to investigate the corrosion which was evident in way of the echo-sounder transducer housing, and discovered that a normal (for wooden yachts) bronze housing had been fitted instead of the nylon-coated one which I had personally delivered to the builders. We knew from other people's experience that if the ship was to be exhibited at the Amsterdam boat show there would be a risk of the topside and bottom plating being faired up with filler, so I had extracted a promise that this would not be done; but on the slip we discovered that filler had been used and that already large areas of it were peeling off. By rights the whole ship should have been sand-blasted and a complete new painting programme carried out, but as we were reluctant to face the very considerable cost we decided against it, and patched up the rusting steel as best we could; but that was a decision which time proved to be unwise.

I have already mentioned that one of the attractions of the design was the modest draught which would permit us to return to some favourite anchorages inside the one-fathom line; but careful measurements made while on the slip showed that instead of the specified 5 feet 9 inches, the draught without any stores or water aboard was 6 feet 3 inches. Perhaps one should be pleased at getting a bit more boat than one has paid for, but we felt we would now have to seek out some other 'hideaway islands in the sun' with a little more depth of water in their harbours.

Our old friend David Jolly is one of the most skilful and resourceful handymen we know, and Susan and I were deeply grateful to him and his wife Susie for giving up many of their summer week-ends to work for us on board. While the two Ss plied sewing-machine and dolly, making awnings and covers, altering upholstery and adapting linen (we like sheets on our bunks, and it was Susie who showed us the dodge of making the bottom sheet with a triangular pocket at each corner to fit the mattress so that it cannot get ruckled up), David and I worked on deck and in the engine-room. We shifted the panel of engine instruments from its wet position at the foot of the steering-wheel pedestal to a dry one within the engine-room hatch; we shortened the staysail boom so that it did not foul the running sail booms, replaced the cockpit light, which was full of water, and rigged up some additional lights below. A casual observer might well suppose us to be highly dependent on electricity, for we used it for weighing anchor, pumping fresh water, ventilating heads and forepeak, lighting—with mounting astonishment I counted 24 lights with 20 switches, and half a dozen plug-in points—and of course for starting the main engine. But apart from this last requirement, there is nothing done electrically that cannot be done, though very much less conveniently, by hand, wind, or paraffin.

David discovered the reason why the auxiliary generator refused to charge after a few minutes' running. In its control box are two diodes (remarkable little electrical one-way valves), and a thermo-switch (a circuit-breaker, which should open in the event of it getting heated by an overload or a short-circuit). When at work the diodes generate considerable heat, and as the heat-sink on which they were mounted was tiny, and the control box in which they were installed was almost airtight, the heat could not escape, so very soon the temperature rose and opened

▶

5. The 4-cylinder Ford diesel dominates the engine-room, which is the full width of the ship and has 4 feet of headroom. The diagonal pipes against the forward bulkhead are cockpit drains. At the top of the lower picture can be seen the manifold of bilge suction cocks, and beneath them one of the two steel battery boxes.

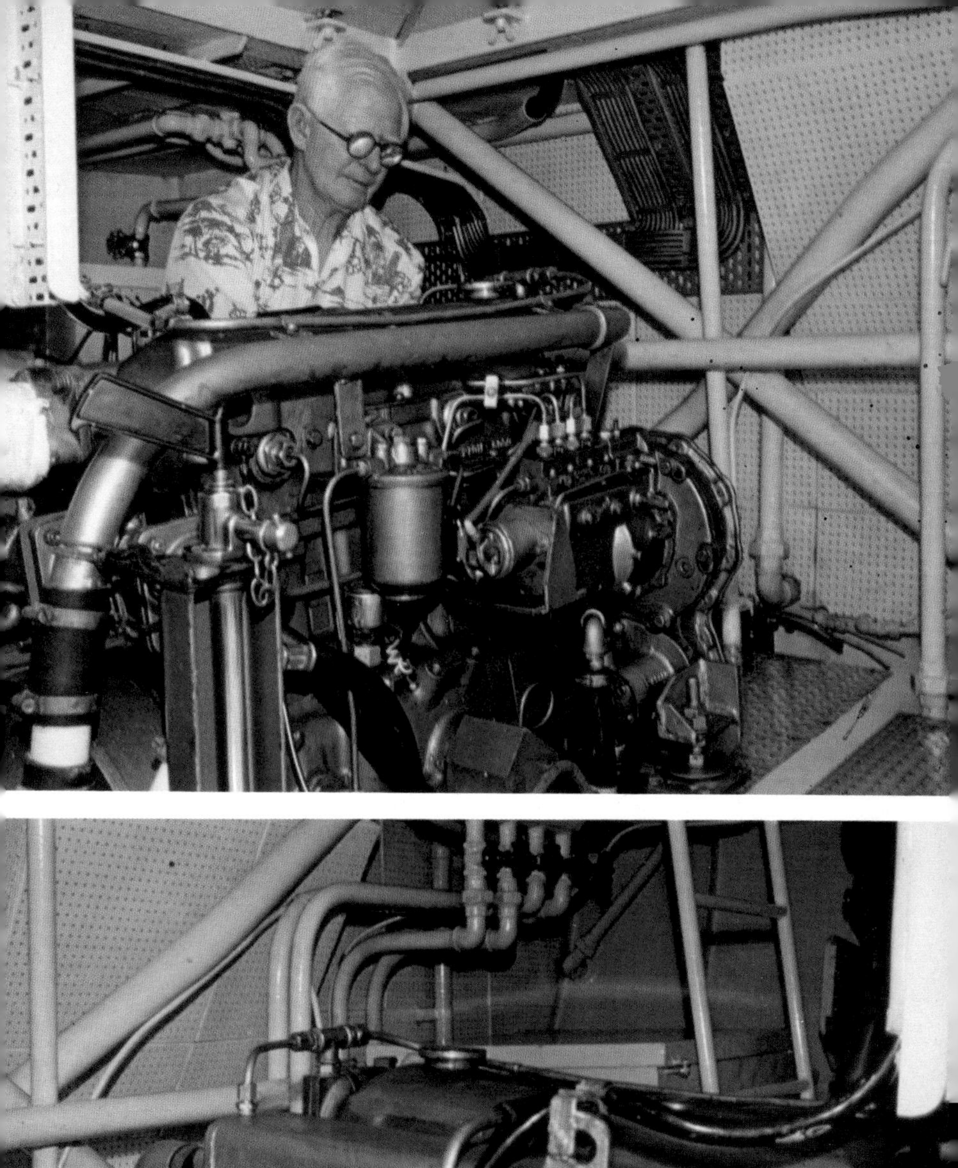

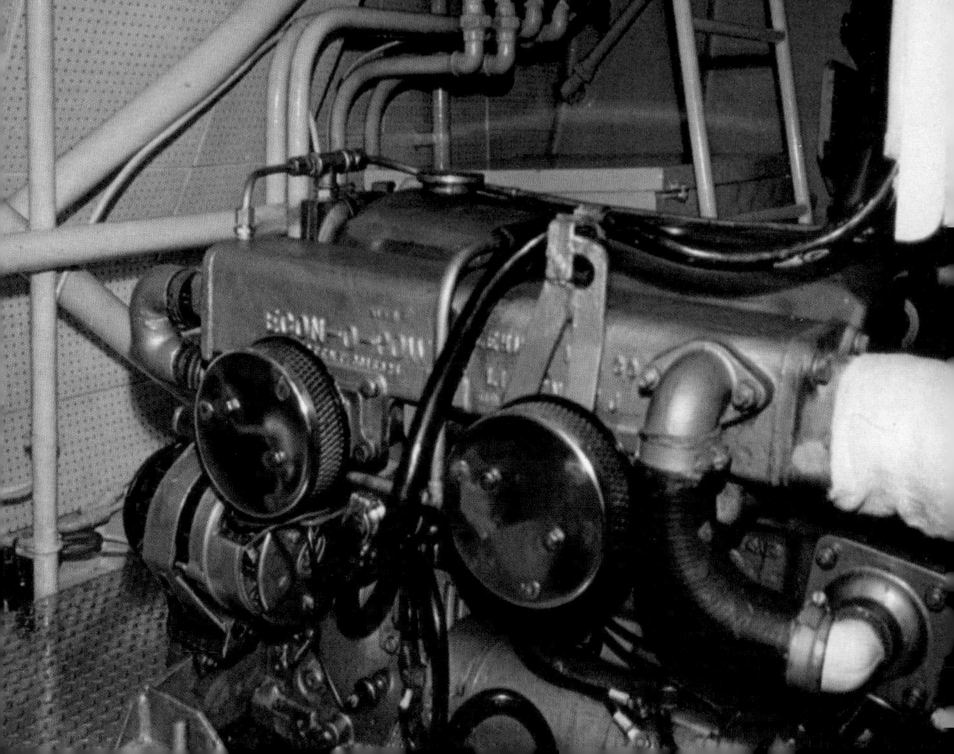

the thermo-switch and no more current could reach the batteries until the whole affair had cooled off. With my new electric drill—we have aboard an American ex-service dynamotor which steps up the ship's 24-volt supply to 240 volts; this also enables me to use normal mains-voltage photographic equipment in the darkroom—we bored 100 ventilation holes in the control box, and we provided the diodes with a larger heat-sink. But all this made no difference, so we replaced the thermo-switch with a hand-operated one, and since then have been able to charge for as long as we wish.

David's only failure was his inability to make the paraffin-operated refrigerator work; although its little lamp burned clean and bright the temperature in the cabinet was not lowered by so much as a single degree. No doubt I ought to have had an expert from the makers examine it, but it did not seem a matter of great importance just then, and there were so many urgent things needing attention that I never got around to it.

As we worked away making alterations and improvements and correcting defects, it sometimes seemed that we would never be ready to leave before the summer was over, and so busy were we, and so often without some vital part of our gear, that we never found an opportunity for trying out our new ship under sail except in light airs, so we still knew very little about her. Although we looked forward with keen excitement to our forthcoming voyage, we were not able to give quite so much thought and attention to its detailed planning as we had in the past. However, we reckoned to get down into the north-east trade wind in October, and cross the Atlantic by that easy, fine-weather route to visit once again the sun-bathed islands of the West Indies where the steel bands play. Early in the new year we intended to sail across the Caribbean and pass through the Panama Canal, but then, instead of continuing to the westward as we had in the past, we would turn into the North Pacific, which we had never visited, and attempt to coast all the way to San

◀

6. Light weather in the north-east trade wind. A running sail with its boom pivoted aloft (an arrangement we were to scrap later) is set as well as the mainsail and mizzen staysail, but the mizzen has been stowed because it was stealing wind from the staysail.

Francisco; however, because of the prevailing headwind and foul current we might have to modify that part of our plan. Hawaii would be new ground for us, too, and from there we would slant down across the equator into the South Pacific, and by way of islands eventually reach New Zealand, a country which we longed to visit again, and this time we would stay longer, cruise round it, and continue to Tasmania. But as we no longer had any ties in England, and would, like the snail, be carrying our home and all our belongings on our backs, the voyaging that we planned to do in *Wanderer IV* could be more leisurely, and therefore perhaps even more enjoyable than the earlier trips in her predecessor, throughout which we had to keep to a fairly rigid timetable. As yet our long-term plan was uncertain, but it seemed likely that we might make New Zealand our base for some time, and from it cruise the South-west Pacific. But, we often reminded ourselves, New Zealand, by the route we planned to take, lay 17,000 miles away, and in so great a distance and the time involved much might happen.

We paid a last visit to the little stone church with the tower where Susan and I had been married 27 years before, and went to the house which had been our home (on and off) for the past 20 years; how green and trim was the grass which we had sown and mown, how thick the hedges on which we had worked with hook and shears, how healthy and richly coloured the roses which Susan had planted and nursed, and how lovely the sunlit views of the downs and the Solent. That evening we dined with friends ashore, and the guests of honour were Miles and Beryl Smeeton and their crew Bob Nance, on the eve of their departure in *Tzu Hang* for Vancouver by way of Cape Horn.

# 2

# To the West Indies and Panama

On the morning of 14 August, deeply laden with stores, water and fuel, we extricated ourselves with the harbourmaster's help from our constricted berth at the piles, and slipped out of Yarmouth's crowded little harbour. Under power, for there was no wind, we made our way a few miles east to spend one last quiet night in unspoilt Newtown Creek, where the only sounds were those of birds, and the trickling caused by the tide leaving the mudbanks; but we remained for five as the weather seemed a bit too boisterous for us to take our untried vessel out to sea. The delay enabled us to attend the goose party which George and Mercia Seabroke gave in the garden of their lovely home overlooking the harbour. Then we sailed to Portland to spend a day in fog, and on to Brixham to spend another there, and so thick was it that we could at times see only our nearest neighbour twenty yards or so away. Finally, in poor visibility we moved round to Dartmouth to have a defective bilge pump dealt with by Phillips.

Our final, misty glimpse of England was of the high ground behind the Prawle, where fields of corn ripe for harvest made pale patches among the green of pasture; we wondered when, if ever, we would see it again. However, we were for the moment far too concerned by a pool of water which was accumulating in the engine-room bilge to feel nostalgic. *Wanderer* has four bilges each, it was said, made watertight from the others by steel bulkheads, and an elaborate system of pipes and cocks connects these to the hand- or power-operated bilge pump. I could not find a leak in the engine-room, and having pumped out the bilge I said I would not look again for two hours; if there was water in it then we had better go to Falmouth for repairs; if there was none we would break the seal on the bonded stores locker and have a drink.

In two hours' time there was more water there, but Susan had a look and reckoned it was seeping up through the cement which covers the lead ballast. I could scarcely believe this, but after a while I had the sense to go and inspect the aft bilge under the sleeping cabin, and there found plenty of water with more trickling in through the propeller-shaft gland. What a relief! A few minutes' work with a couple of spanners fixed that; but so much for our 'watertight' bulkheads! We had our drink, and that so fortified us that we then inspected the other bilges, and found that what we call the 'hold', the big space under galley and chart-room and lying between the fuel and water tanks, in which we stow cases of canned food, also had some water in it. The source of this we traced to a faulty joint in the freshwater system, and we cured it with Araldite glue.

It may have been realized by now that I came new and rather reluctantly to steel construction. Certain things have been impressed on me as being important, and one of these is that there should never be any water below decks. I do see the force of this, having regard to the buckets full of rubbish—wood shavings and sawdust, insulating material, brass and steel screws, and horrors! bits of copper wire, for these as I noted earlier are death to steel if water is around—which we have fished out from under the sole, and the knowledge that there is plenty more which we cannot reach because it has been built in.

So I mopped out that bilge, too, and with a fair wind we stood on to give Ushant a berth, and soon were running under the mainsail only, for we had quickly discovered what I had previously suspected—the mizzen of a ketch is a nuisance when running for it interferes with the wind into the mainsail and makes that sail restless and inclined to chafe itself on the rigging. We had a pair of twin running sails, one of which would have been a help then, but with so many other things to attend to during the past three months I had omitted to reeve their gear.

The crossing of the Bay of Biscay was uneventful, except that for the whole of it visibility was poor, and for much of the time there was fog; on one occasion only did it clear sufficiently for me to fix our position by observations of the sun. For the rest of the time low cloud obscured the sky, and the few ships which passed within our small circle of visibility were phantom shapes, as grey as the mist itself.

We steered for Cape Ortegal, intending to put in either at Cedeira or El Ferrol; but the visibility continued to be so poor as we approached the Spanish coast with a fresh onshore wind that we did not care to close with it, uncertain of our position as we were. So we reefed the mainsail to slow us down, and steered a course parallel to the shore, hoping in time to get some observations, or a clearing of the weather. We got neither, but next day did get a rather uncertain D/F bearing of Cape Finisterre. That evening, however, the fog grew so thick that we did not fancy crossing the shipping lane, to the westward of which we now lay, so under the reefed mainsail only we hove-to for the night. This was an experiment, for although our earlier yachts managed very well like that, it was questionable whether a ketch, with her mainmast so far forward, would behave properly. We found that she lay very quietly, but instead of the wind being about 45° off the bow (which I think it should be) it was more nearly abeam, and although that did not matter at all in these gentle circumstances, we did wonder what would happen when we had to heave-to because of stress of weather. We turned on all our navigation lights and had a wonderful sleep.

At daybreak we made all plain sail and by D/F radio homed on Cape Finisterre, which at noon on our fifth day out from Dartmouth loomed out of the mist ahead. As soon as we had rounded the cape we left the fog astern, and in sunshine had a glorious sail hard on the wind, but still a little puzzled by our ship's desire to turn to port, to the village of Corcubion, which had been much tidied up since last we were there. After one night at anchor we punched on against a fresh southerly, and in mist steered a succession of compass courses to circumnavigate the several unmarked offshore dangers—the Bajos Mexidos, Correbedo, and Las Baleas—without so much as a glimpse of a mountain to check the dead reckoning, and I was astonished and much relieved when the striped lighthouse on the south end of Isla Salvora materialized ahead at the predicted time. There we entered Arosa Bay, and in the evening let the anchor go off the fishing village of San Julien, while the wind blew, the rain came down, and the fiesta rockets went up. But after five fathoms of chain had run out it jammed in the pipe leading from the chain locker, and we found ourselves in an awkward situation with the anchor on the bottom but with insufficient chain to let it get a hold, and all around us the moored mussel rafts. Of course we let go the kedge on its nylon warp, which within a few minutes had turned yellow, due to some form of pollution in the water, and after much hammering and twisting we managed to free the chain.

As a bower anchor I had specified a CQR. (the original, genuine plough type) of 60 lb, and for cable a 45-fathom length of ½-inch Lloyds-tested chain. To handle this I chose an electric windlass from the same firm that supplied the chain so as to ensure that the chain and gipsy were in harmony. The windlass went to Holland to be fitted, but to save freight charges the chain remained in England to be picked up on our arrival there. On a visit to Holland after the windlass had been installed I did think the navel pipe provided by the windlass-makers looked rather small to take the big end-link of the chain, but I knew how easily one can be deceived, especially in a vessel larger than one is used to. So back at Yarmouth I made a replica of the big link in wood, and when we went to bring the vessel home I offered this up and found it would not pass through the pipe. In a wooden vessel one could easily fit a larger pipe, but *Wanderer*'s Builder had, and in my opinion rightly, provided a pipe of similar size leading to the chain locker, and this he had welded to the steel deck; as the deck was covered with plywood and teak planking was laid over that, he was naturally reluctant to do anything about it. So when we came to ship the chain, the first thing we had to do was cut off the big link from the bitter end. It has always been our habit to end-for-end the chain every year as this keeps it supple and doubles its life, but this we now can never do as it is impossible to attach a CQR. anchor to a short link without using a shackle of too small a size.

As originally fitted, the chain pipe led vertically down from the windlass to the narrowest part of the huge chain locker, so that the chain could neither run in freely nor clear itself when running out. With a hacksaw I therefore cut out from the pipe three narrow segments so that I could bend the pipe a little and lead the chain farther aft to where the locker is much wider. This had worked well enough until we came to anchor off San Julien, when a twisted link jammed in the bent part of the pipe and we found ourselves in the awkward situation I have already described. We could not risk such an occurrence again, so I spent the next two days removing all of the pipe except the few inches where it passed through the deck, and building in its place an open-fronted wooden chute shod with metal at the places where the chafe of the chain would be greatest.

Written at sea,
8 November 1968

I have just washed myself all over in fresh water; Susan has done the same and is now busy with the laundry, for it is a fine drying day. During the night Nicholson devoured a couple of quite large flying-fish, judging by the size of the wings he left lying on the saloon carpet, and now with twitching tail and a smile on his lean Antiguan face is sleeping them off. Such celebrations are, no doubt, due to the fact that we shall at noon today, in latitude 17°20′N., longitude 36°32′W., be just past the half-way point of our passage between Gran Canaria and Barbados. To reach this point has taken longer than it should, 16 days, whereas the much smaller *Wanderer III* once did it in 12. But this year something seems amiss with the north-east trade wind—we and others have known this happen before. So far it has blown with little strength, and as I sit typing in the saloon, I can tell by the easy motion and lack of noise that we are doing no more than 3 knots.

We have been looking forward, perhaps a little apprehensively, to the first tough weather in our new ship, but so far calms and light airs have been her portion. We were in Spain for a month, but did not want to move about much as we still had a lot of work to do on board. I am amazed at how many things I *can* do (though not always efficiently) if I have to. In turn, ably asssisted by Susan and the occasional visitor, I have played carpenter, rigger, plumber, electrician, engineer; I have

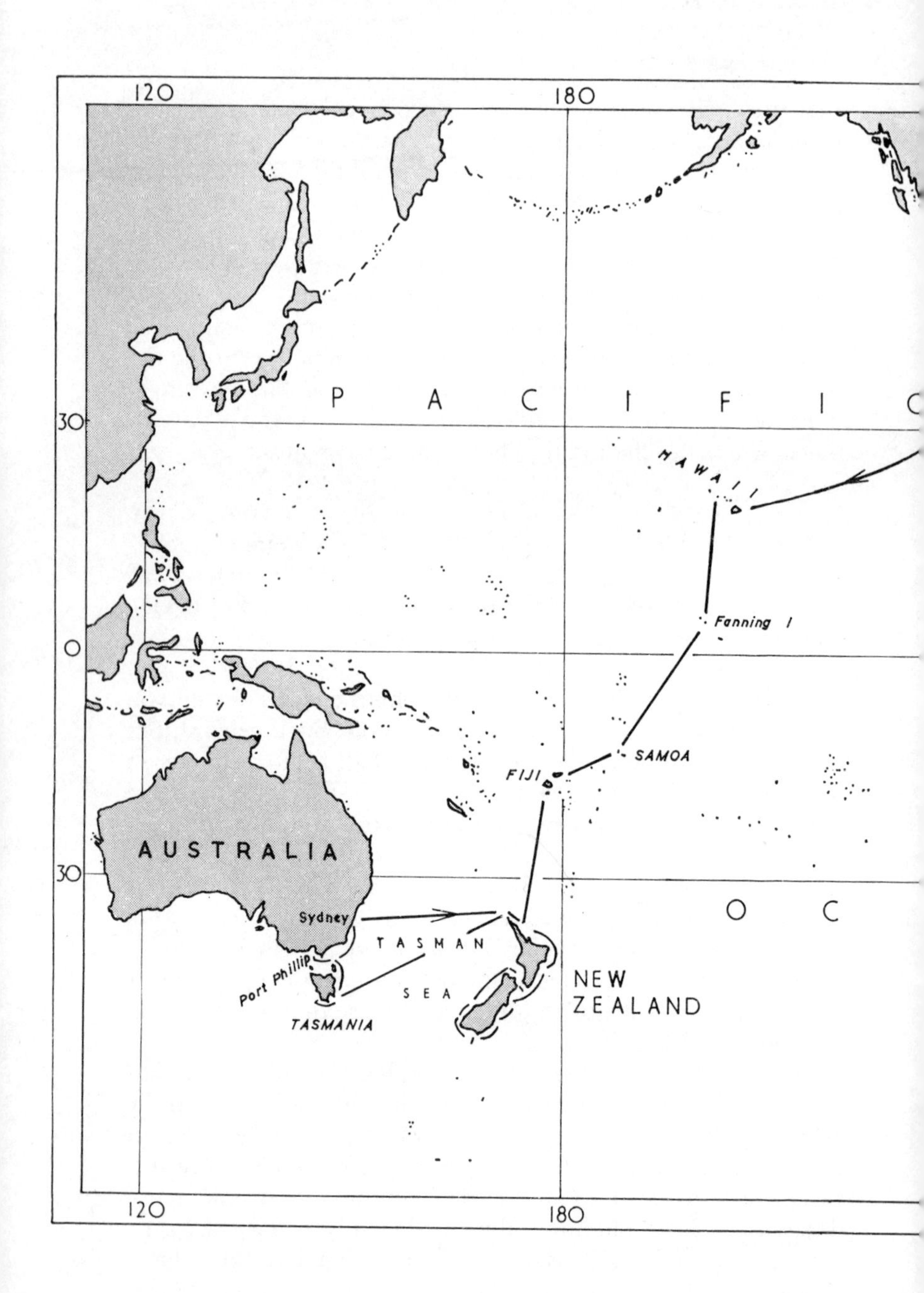

120
180
P A C I F I C
30
HAWAII
Fanning I
0
SAMOA
FIJI
AUSTRALIA
30
Sydney
O C
TASMAN
SEA
Port Phillip
TASMANIA
NEW
ZEALAND
120
180

NORTH
AMERICA
SOUTH
AMERICA
AFRICA
ATLANTIC
OCEAN
AN
San Diego
ancisco
W. INDIES
Bonaire
Barbados
Cristobal
Madeira
CANARY Is.
Bayona
60
30
0
30
60

even become familiar with the serpentine insides of a gas water-heater, but failed to make it work. So far I have not learnt to weld, but often wish I had.

Our time in Spain was shared between the upper part of Arosa Bay and Bayona, where we rode out a week of wild weather, the same that caused England's worst floods in 15 years. We gathered so much rain in the dinghy that we repeatedly washed ourselves (you will think this is becoming a fetish) and our clothes without having to draw water from our tanks. This was just as well, for our waterworks had broken down, the electric pump which feeds the pressure system having gone out of business, leaving us with a small and inefficient hand-pump in the galley. For two days Susan and I wrestled with it; we cleaned out an astonishing quantity of dirt and made new gaskets, but with no improvement. Also we were worried by the steering, which had become so stiff that two hands were required to force the wheel over, and I feared that so much strain would damage the Mathway steering gear—a sophisticated set-up of bevel boxes, torsion tubes, and universal joints, which I was unable to disconnect—and would quite exhaust the helmsman.

Then one day John Rock arrived, single-handed in *Zest of Dorset*. I called on him to invite him over for a meal, and was glad to learn that by profession he was an engineer. When I told him of our troubles he kindly offered to assist. With one judicious blow with a hammer on the end of the drag-link he disconnected the Mathway gear, and we at once saw that there was nothing the matter with that excellent affair. Investigation showed that the upper part of the steel rudder stock was held in a nylon-bushed bearing just below the screw deck-plate where an auxiliary tiller can be shipped; rain had entered at the plate, got into the bearing, which has no proper means of lubrication, and rusted up the stock. John poured in some freeing oil, and after leaving this for several hours to penetrate, we shipped the auxiliary tiller and together forced it slowly to and fro. Intermittently we kept this up for 24 hours, and when the freeing oil showed signs of seeping through, we replaced it with a mixture of lubricating oil and paraffin, and finally with oil alone; in this manner we eventually got the rudder to move freely, and reconnected the steering gear. Our engineer friend then tackled the fresh water pump, found that one of its rotors had become displaced on the shaft, and soon had it in working order. We pumped out the freshwater-tanks in the hope of ridding them of some of the sediment they contained, and refilled them by jerrican with good water brought off 18 gallons at a time from the friendly yacht club. This gave us the opportunity to measure the capacity of our two tanks, and mark the dipstick in gallons instead of litres. To our astonishment we found that instead of the 300 gallons they were said to hold, the total capacity of the two tanks was only about 200 gallons.

On 27 September we left Bayona with regret, for we regard it as one of the best anchorages in Spain, clean and quiet as it is, and with the pleasant little town near-by for stores; also, now that the magnificent ramparts of Monte Real, the great fortress which flanks the bay on its western side, are floodlit, by night it has an enchanted, fairy-tale appearance. We were tempted to go to Cascais in Portugal to visit friends, but the recent bad weather suggested that it was now getting late in the season, and the open anchorage there might not be tenable; so we headed direct for Madeira, 700 miles distant.

On the passage to Spain we had grown to dislike the staysail boom for which I had asked. It was a heavy spar which crashed about frighteningly when the wind was light, and then was a danger to anyone working on the foredeck; on one tack it prevented the forehatch from being opened fully, and it had to be topped up and secured before the sail could be lowered. So while in Spain we had unshipped it and made arrangements to set the sail boomless. On each sidedeck about 4 feet abaft the main rigging there is a pair of bollards put there to take the springs when lying alongside, and the forward of each of these was in the right fore-and-aft position to take the new staysail sheets. Each sheet started at its bollard, went forward through a block shackled to the clew of the sail, then back through a fairlead block on the same bollard and thence to a cleat on the cockpit coaming. On the passage to Madeira we were able for a short time to try out this new arrangement, and found that the single whip purchase of 2 to 1 was not sufficiently powerful to sheet the sail in hard enough in anything of a breeze. So I made up a small handy-billy to give an additional purchase of 4 to 1 when required, a seamanlike piece of gear which is easy to rig and very effective, and, as we had some old rope-stropped blocks on board, cost us nothing, whereas a pair of winches to do the same job would have cost about £200. We gave the boom with its stainless fittings to an astonished and delighted boatman at Funchal, and were a happier ship ever after.

For most of the trip to Madeira the wind was fair, but towards the end it fell right away and we made considerable use of the engine, much to Nicholson's disgust, for he hated the whine of the shaft, which presumably was on his wavelength, and found it necessary to move from his quarters in our sleeping cabin to a berth in the saloon. But we noticed as time went by that he tended to use the saloon more and more, even when we were under sail, and eventually rarely went aft except to use his heads; as cats know where the comfort lies, we found ourselves following his example, for although in quiet weather the cabin was comfortable and quiet, and one was not disturbed there by the watchkeeper who might want to get coffee, look at a chart, or obtain a radio time signal, in rough weather the motion was more violent than it was amidships; also as the bunks were not parallel to the fore-and-aft line, ones head was lower than ones feet when using the lee one.

At dawn on our fifth day at sea, when we were close to Madeira's eastern point, we stopped to watch the rising sun paint the mountain slopes with changing colours, and to have breakfast; then after a general tidy-up we motored on, entered the harbour of Funchal before noon, and were berthed by a pilot close to the Cais do Cidade, with our anchor out to the south and our stern secured to a mooring, among the other visiting yachts which were also moored fore-and-aft. This is a great improvement on the old arrangement when yachts lay to single anchors, for one is head to the swell, which sometimes comes rolling into the harbour, and there is now little risk of collision. I thought we were all much in the way of Blandy's smart, varnished launches carrying passengers to and from the succession of tourist ships which called, but their coxswains always had a wave and a cheerful word for us. Joao Borges—he owned the jewellers shop called Big Ben, but in marked contrast was a diver of considerable accomplishment and courage, and had the previous year, almost single-handed, salvaged a sunken fishing-vessel with two Rolls Royce engines and was using her for his own purposes—came to invite us to make use of the landing and the showers at the Club Naval, and to extend to us, as he did to all visiting yachts no matter what their nationality might be, much kindness and hospitality.

When Susan and I started ocean voyaging in the early 1950s there were not a lot of other yachts doing the same thing, and in the autumn in a port such as Funchal there might be one or perhaps two others about to set out to the westward. But today there are many more voyaging yachts, and a crossing of the North Atlantic is almost a commonplace. During our stay of two weeks there were fourteen visiting yachts, mostly British, and all were Atlantic candidates; they ranged in size from the tiny Westerly *Eeyore*, sailed by Nigel Church and Bill Woolley, two young merchant navy officers, to H.M.Y. *Britannia*, in for bunkers on her way to embark the Queen in South America. Our last night in port we dined with our friends Alec and Yvonne Zino—he owned the uninhabited island of Selvagem Grand, half-way between the Madeiras and the Canaries, and on it had built a house which he claimed was proof against marauding Spanish fishermen—in their cool and gracious home on the hillside, together with some of the officers from *Britannia*.

I was seated next to Admiral Morgan, and must have been enthusing too much about *Wanderer*, for towards the end of that splendid meal he turned to our hosts and said:

'I think it's time we went and had a look at a *real* yacht.'

So we all drove down to the harbour and the party continued very pleasantly aboard *Britannia*. When Susan and I made our way down the brightly lit gangway late that night, we declined the lift offered by our hosts, and arm in arm took our last walk along Funchal's familiar waterfront.

The Admiral had bidden us to go along and salute before leaving, so next morning we made our way up the harbour and dipped our biggest ensign, and having received in return a heartening send-off we put to sea. We learnt later that Alec Zino had, as it were, thumbed a lift, and got a passage to his island aboard the royal yacht.

Ours was an easy trip with a light, fair wind and no incident, and two nights later we crept cautiously into the light-reflecting oily water of the harbour of La Luz in Gran Canaria, where much of the anchorage was already filled by a mass of shipping which daylight revealed as a brood of Japanese fishing-vessels clustered round their mother-ship. After our last visit we had said we would never go there again because of the abominable filth, but the availability of cheap and excellent provisions after very expensive Madeira, where only the wine was cheap, proved too strong an attraction. We got all our needs at Rodriguez's store, which put the lot together with ourselves into a truck and delivered us back at the Real Club Nautico, where we had left our dinghy. No doubt the club has in its time seen too many potential Atlantic voyagers, and no longer offers them temporary honorary membership. Short though our stay had been it was long enough for our topsides to become coated with the black oil for which the place is famous, and although we removed this just before leaving, it left a yellow stain on the paint. But I have just been up in the brilliant sunlight and had a look over the side and am glad to see that the stain has almost been bleached out.

We had been at sea for a week when there occurred something which since the end of the great days of sail had grown rare, but now with the increasing number of ocean-going yachts is becoming more common again—we sighted a sail ahead. As there seemed no possibility of overhauling this before nightfall, we started the engine and eventually came up with the cutter *Waltzing Matilda*. She had been built in Tasmania in 1949 for our Australian friend Phil Davenport, who had made a great voyage in her to England by way of the Straits of Magellan. Now, under American ownership, she was bound towards Barbados like ourselves, and was 10 days out of Santa Cruz de la Palma. Having spoken her we motored well clear for the night, and have not used the engine since. In the morning *Waltzing Matilda* was still in sight astern, but no longer under twins; instead she carried a press of sail, and to avoid being overtaken by her we set for the first time our 400-square-foot mizzen staysail; with the wind on the beam this pulled well, and slowly we drew away from our consort; we have not seen her again. We found the mizzen staysail rather limited in use, working really well only with the wind from abeam to about two points abaft the beam, and we began to understand what Michael Handford of Cranfields had meant when I asked him to have it made.

'I do not know', he wrote, 'whether you should regard this as a luxury or a necessity.' Certainly we have found it fun to play with on fine days.

Incidentally, we continue to find the mizzen a somewhat mixed blessing, and in common with most owners of ocean-going ketches we tend to make little use of it and treat the ship as a cutter, though the loss of its 200 square-feet certainly is a handicap. Except in moderate winds, when we are close-hauled the mizzen gets back-winded by the mainsail, and although this could probably be overcome by increasing the gap between it and the main, I doubt if the necessary reduction of mainsail area would be a good thing. When the wind is aft we usually take the mizzen in to stop it making the mainsail restless, or gybing itself unasked; on a broad reach it certainly pulls well, but in a squall tends to induce considerable weather helm.

Written at English Harbour,
3 January 1969

A steel band plays at the Inn, its thrilling beat and melody wafting out to us in gusts on the warm night air. When it pauses for a rest, the music of cicadas and the bell-like notes of the tree-frogs take over. The massive stone pillars of the old, roofless boathouse are softly lighted; astern of us there are lights and silhouetted figures at the Powder Magazine, the home of Vernon and Emmy Nicholson; and a hundred yards away at the head of the harbour a light shines from the House Beyond, where Ann Worth is in residence, and her *Blue Shark* lies at the pier. When Susan and I first came here in 1952, there were, including our own, only six yachts; now there must be at least thirty, and the harsh glare of their crosstree- and flood-lights does not, I think add to the night-time attraction of the old dockyard round which they are clustered. But the moon is just rising over the low hills to windward and will soon put the man-made lights to shame.

The second half of the Atlantic crossing was an improvement on the first, for the wind strengthened, though never enough to give us a day's run of more than 150 miles, and that on three occasions only. There was only one brief shower of rain, and we became quite anxious for a small, blue sloop which, with man and wife aboard, had left La Luz the same day as ourselves, for she carried only 20 gallons of water, and although this should be just enough for two people for four weeks, she would almost certainly take longer than that over the crossing.

Ever since we took our new ship from Holland we have been bothered by her tendency to turn always to port. This is particularly marked when under power, even though she has a left-handed propeller, which of course should turn her the other way. We had so little opportunity for checking it when under sail in a decent breeze, that it was not until the trade wind run that this vice became really obvious. We have a pair of twin sails for self-steering when running before the wind, but have found no advantage (except lack of chafe) in using them, for as soon as the wheel was left free the rudder started to turn to port and, if not corrected, firmly turned the ship to port and back-winded the port sail. If the mainsail was set, and was out on the port side, the ship still turned that way and a gybe resulted. Constant care had therefore to be taken by the helmsman when running, and we soon became tired with the effort required.

The booms for the twin sails are hung permanently aloft, their goosenecks being at the root of the lower crosstrees, and the act of setting one of these sails, each of which is 270 square feet in area, lifts its boom to a nearly horizontal position. When not in use the lower ends of the booms are lashed to the main shrouds. We have found that this arrangement is not very satisfactory. For one thing a boom is not under control while its sail is being set or handed, and although made of light alloy, swings about in a highly dangerous manner if the ship is rolling; for another, because the pull of the after guy leads down at a steep angle, it places a heavy strain on the leech of the sail. With this in mind I had asked for the leeches to be strengthened with wire, and this had been done; nevertheless we carried away one of the wires and damaged its sail early on the crossing. We did

▶

7. At English Harbour, Antigua, Susan spread out her Vivitex in the cool cloisters of the old spar and lumber store and made a cockpit awning.

KLEENEX

not use the pair of them much because, as I have said, the ship would not steer herself under that rig. However, we often needed one of them when running to increase the total sail area, and to help balance the mainsail (Plate 6), but although we experimented we failed to find any way of controlling the boom during sail-handling, or of reducing the heavy loads involved. We do not like carrying one of these sails in a wind much greater than 20 knots, so when the wind shows signs of freshening we become tense, and cannot relax until we have taken it in and tamed its wildly swinging boom, which is a two-man job, during which there is a risk that one or other of us may get injured, or damage be done to some part of the gear.

We had installed in England an electric automatic helmsman called a Pinta (someone had remembered the mysterious ghostly helmsman who, Slocum claimed, steered the *Spray* at times) and this was of such great value when we were both needed on the foredeck together for sail handling, or to enable the watch-keeper to brew a cup of coffee or take a sight, that we were reluctant to use it more than necessary for fear it would wear itself out, which we felt it might do because of its determination to move the wheel between 25 and 35 times a minute. When I mentioned this to the skipper of *Eeyore* at Madeira, he pointed out that in merchant ships auto-helmsmen go on working without trouble or attention for years, and there was no reason why ours should not do the same. Day by day as our line of noon-position crosses crept slowly across the chart of the North Atlantic, we found ourselves switching on the Pinta more and more often, until eventually we had it working almost all the time. Its consumption of electricity is so small (the makers claim an average of 25 watts, i.e. about 1 amp on our 24-volt system) that we needed to run the auxiliary generator only once a week for between 5 and 7 hours, and this also looked after all our other electrical requirements, including navigation lights. Once we had come around to making full use of the Pinta our tradewind run became much more easy, comfortable, and enjoyable.

◀

8. Sorting and stowing stores at St. George's, Grenada, under the new awning. At least one advantage of our big ship now became obvious—there was no lack of space.

Nevertheless the steering bias to port is disturbing, and since our arrival in the West Indies several knowledgable people have offered suggestions for its cure. It would seem that the sides of the rudder, which is a double plate affair of streamline section, are not identical. It is unlikely that we will find nearer than the U.S.A. a steel-worker with sufficient skill to fair it by the local application of heat, but if the depression or flat is sufficiently obvious, we might be able to build it up with epoxy glue ourselves. Another suggestion is that we have a trim-tab welded to the rudder's trailing edge, but if this is permanently angled to correct the bias when running, it would presumably increase the already considerable weather helm which the ship carries when reaching in fresh winds under full sail with the wind on the port side. We hope instead to follow a third suggestion; this is to fit two wooden, wedge-shaped 'spoilers' to the after end of the rudder in an attempt to kill its hydrodynamic properties and make it 'weathercock'.

The 2,777-mile crossing to Barbados took us a few hours under 27 days, which is certainly no record. We anchored in 12 fathoms in Carlisle Bay, a long way out because of the small-boat moorings with which the best positions are filled, and rolled heavily. Alas! the Aquatic Club, which has in its time welcomed so many Atlantic voyagers and provided them with a safe place at which to land and leave the dinghy, is no more; a multi-storey hotel has been built on its site. The surf on the beach made landing awkward, so as we now have a Seagull outboard motor, we went for shopping by dinghy to the sheltered but crowded Careenage. Although we never again saw *Waltzing Matilda* during the ocean crossing, she must have been close to us for she reached Barbados only 12 hours after we did. A little while later she was lost while sailing between the islands of St. Lucia and Martinique. As I understand it, she was reaching, but with the booms for her twin running sails still rigged out in the horizontal position; in the strong wind between the islands she heeled suddenly and rather far, and dipped the lee boom into the sea, which drove it aft with such force that it carried away the lee rigging and broke the mast. A ship went to her assistance, a member of her crew was badly injured, she was abandoned and was never seen again.

For us two days of rolling were enough (we feel the bright, sugar-growing island of Barbados has now lost most of its attraction for a voyaging yacht) so we ran overnight to Bequia, and then made our way down the chain of Grenadines, which normally provide such excellent sailing but this time produced only light airs, to Grenada, where a lot of mail and a cantankerous customs officer awaited us.

At Madeira we had need for the first time of our absorption type, paraffin-burning refrigerator—a luxury we had never enjoyed before, and around which the galley had largely been planned. As on an earlier trial, it gave off some heat but no suggestion of cold. But in Grenada they understand these things, for many are still in use on the island, and the makers have an agent there. He found that the sinuous works of ours had been damaged and all the refrigerant had escaped, and as it could not be refilled there was no alternative but to buy for £20 a complete new set of works. The dismantling of the joinery-work, for the unit had been built in with the apparent intention of excluding the air which it needs if it is to function properly, the removal of the old works and the fitting of the new, and the proper ventilation of it, provided me with four day's work in a locker measuring 3 feet by 2 feet by 2 feet, while the thermometer stood at 85°. I bored plenty of ventilation holes, and now we enjoy the luxury of cold food all for less than a gallon of paraffin a week, and we can have cubes of ice in our evening drinks. American visitors, who usually choose 'scotch on the rocks', often ask us how many hours a day we have to run a generator to keep the refrigerator going. When we reply that we don't, for this is kerosene ice made without a sound, some of them actually smell it.

In the lagoon, to which a channel had been dredged since our last visit, we found Grenada Yacht Services firmly established and offering all facilities including a marina and synchro-lift. We booked a berth on the latter for early February so as to scrub and paint our bottom, which was growing goose barnacles and patches of something that looked like coral, and to see what could be done about our rudder; then we headed north for 300 miles up the chain of islands to Antigua. We made the usual stops, found the usual calm patches and windy passages, and eventually sailed swiftly into English Harbour in time for a round of Christmas parties. Here we are attending to as much of our refit as possible—the list of what we have done so far fills one and a half pages in the log-book—and in the cool cloisters of the now repaired spar and lumber store Susan spread out a length of Vivitex (an American, proofed cotton material which does not readily mildew) and made a tropical awning (Plate 7) to replace the earlier one of light terylene, which let through too much light and rattled in anything of a breeze. It is a great improvement, and after she had fitted it with side curtains which can be rolled up or lowered to guard against the low morning or evening sun, we spread it and since then have been enjoying all our meals out in the now shady and much cooler cockpit, which with thick, white cushions on the seats, the scrubbed teak deck underfoot, and at evening a shaded light swinging gently overhead in the breeze, is glamorous and quite the equal of anything the charter fleet can offer.

But our remarkable ship continues to provide us with surprises. One day, while stowing cases of canned food in the hold, Susan noticed that the skin fitting which holds the echo-sounder transducer was leaking, and when I tried to tighten its bolts I found that all of them were broken, so that nothing but bedding compound and water pressure held the fitting from dropping off. With the help of John Guthrie, skipper of *Peloha*, who turned out unexpectedly to be an experienced diver, we have patched it up temporarily. Then one evening just before we were due to go ashore to a party in the dockyard, I in my best shirt and shorts was standing in the cockpit minding my own business while Susan down below was finishing a letter she wanted to post, when a whirring and click of pawls coming from the foredeck caught my ear. I went to investigate and found that the electric windlass had switched itself on and was busily heaving in the anchor chain. I can well imagine what might have happened in this anchorage crowded with expensive yachts that windy evening had this occurred a few minutes later when we would both have been ashore.

Such gremlins as these certainly needed investigating, and rumour had it that the owner of the motor yacht *Grampus* was an electronics expert, but not an easy man to approach. I took my courage in both hands and asked for his help, which was readily given, and within the hour he, Henry Chatfield, came aboard with a highly sophisticated meter. He checked us for electrical leakage and galvanic action, and to our relief pronounced us to be free of both. The failure of the skin fitting bolts he put down to inferior material and fatigue, and after unbolting the windlass from the deck and capsizing it, he discovered that its motor was of the dangerous (for marine use) single wire and earth return type; a drop of water trickling down inside the casing had earthed the current and set the motor working. Now, of course, we keep the master switch, which we have fitted, turned off. How fortunate we have been first to meet the engineer John Rock in Spain just when we needed him, and now to have the assistance of skilful Henry. He has since done some other things to sort out our electrical and mechanical problems, and just as though all this was not enough, he and his wife invited us to a fine turkey dinner aboard *Grampus*, and then made us a present of two gallons of excellent Martinique wine. How, we wonder, can we properly show our appreciation to such very generous Americans.

There are some lovely ships lying here, but the only one I really covet is the *Lord Jim*—we watched fascinated as this magnificent schooner, superbly handled, came in and berthed stern-to at the dockyard under sail—but *Lord Jim* is much too big for us. To my eye none of the rest is quite so handsome as our own, and Susan and I feel sure that if we keep on working hard enough and long enough we will in time persuade her to be a good-mannered and seaworthy ship of which we can feel proud.

In the third week of January we headed south to keep our appointment at Grenada. This was the sixth time we had cruised along the gently curving chain of Leeward and Windward Islands, which are just as lovely as they ever were; with the boisterous passages between them, where the trade wind blows unhampered, the sea runs free, and the spray flies, interspersed by the quiet areas in the lee of the high, green mountains, they provide one of the least demanding and highly enjoyable of cruising grounds. But this time the islanders did not always seem so friendly and cheerful as we remembered them in the past, and this was particularly noticeable in the islands which had recently got their independence, where discourtesy to the white visitor at times seemed almost studied. Some of the shop assistants in Castries, for example, preferred not to sell us what we wanted even though it was on display, and Antigua's new flag, which one was expected to fly as a courtesy ensign, depicts a rising sun on a black, expanding sky above a small diminishing triangle of white. However, the people on Bequia, still British, appeared unchanged; but no doubt airstrips make a difference as they bring too many visitors, and Bequia still had none though one was threatened.

On the appointed day *Wanderer* was lifted out of the water on Grenada Yacht Services' synchro-lift, the like of which we had not seen before. It consisted of a 90- by-25-foot platform lying submerged between two concrete walls, and was supported at each corner by a steel cable leading to its own electrically-operated windlass situated on the wall; the four motors were synchronized; the lift could take a vessel of 250 tons. A warping buoy to windward helped us get *Wanderer* stern-first to the lift, where she was correctly positioned by the yard hands, who cunningly made use of each puff of wind, and a diver then went down to see that all was well. Several vertical supports, controlled in pairs by wires and small winches, were moved in to grip our topsides with their padded arms, the synchronized motors were started, and slowly the platform with us sitting on it was lifted horizontally out of the water. A synchro-lift is a costly piece of equipment, but nothing, I imagine, could be simpler to operate, and certainly we had never experienced such an efficient, clean, and painless haul-out.

Our intention had been to remain on it only for a day, which should allow plenty of time for scrubbing and painting the bottom, and for fitting to the rudder the spoilers which we hoped might cure the ship's strong tendency to turn to port both when sailing and when under power. But when we were clear of the water we found that in several places areas of the filler with which the builders had plastered the hull had peeled off, and two of these areas extended up above the waterline. The steel plating in those places therefore had to be cleaned off and given several coats of paint before antifouling, with the result that we had to remain on the lift for four days. The only person to be glad about this was Nicholson, and once he had discovered the best way of negotiating the steep, wide-runged ladder that led to our deck, he spent most of each night ashore huntin', and sometimes brought his trophies back for display or dismemberment on board.

Now that we knew what to look for the cause of the steering defect was clear; the starboard side of the 'streamline' rudder had a slightly more pronounced curve than that of the port side. By fitting spoilers we hoped to kill the rudder's hydro-dynamic behaviour and make it weathercock, and by making the port spoiler a little thicker than the other, which was at its after end to be 2 inches thick, we thought we might perhaps cure the bias. The spoilers, as specified by Henry Chatfield, consisted of two wooden wedges; each measured 8 inches fore-and-aft and 16 inches vertically, and they projected 3 inches abaft the rudder's trailing edge, where they were bolted to one another, thus gripping the rudder blade. To stop them from sliding off a strip of wood was glued to each side of the rudder, and slots were cut in the wedges to engage with these. With this arrangement we could, if necessary, dive down, and by undoing two bolts free the spoilers for alteration.

G.Y.S. had a rule that owners and their crews might not do any work below the waterline when a yacht was out of the water, so Susan and I could only stand around and watch the spoilers being fitted, and the cheerful painting gang slapping on the paint more and more slowly as the day drew to its close. When one of them spilt some paint over his neighbour's bare foot, one of his mates remarked:

'If you is de boss-painterman, boss, you ain't much good as de boss-painterman.'

But ashore a grisly scene was being enacted.

Because of the recent increase in theft from yachts, nobody except owners and their friends was allowed into the marina without a pass, and a retired policeman, a black, had been engaged to act as watchman at the gate. The young diver who had made sure that *Wanderer*'s keel lay fairly on the blocks, came that afternoon and tried to enter the marina, but as there was no immediate employment for him he had not been issued with a pass, so the ex-policeman stopped him. A slight scuffle ensued, but he was not allowed past the gate. He then walked a mile to the shack where he lived with his mother, picked up his underwater spear-gun, returned to the gate, and in front of several onlookers shot the ex-policeman and killed him.

That evening, under cover of darkness Susan and I refastened the echo-sounder fitting with new bolts, and with paint touched up a few of the holidays the boss-painterman had left.

Afloat once more, and without a scratch or an awkward moment, we at once tried out the spoilers, and much to our relief found that they had cured the tendency to turn to port when motoring; indeed, there was now a slight bias the other way, just as there should be with a left-handed propeller. But alas! before we had a chance to try them out under sail, the special underwater epoxy glue holding the wooden strips to the rudder let go, and we lost the whole bag of tricks. The spoilers should have floated, but we failed to find them, and unfortunate though the loss had been I could not conceal from Susan a quite misplaced sense of relief that even the experts cannot make glue work, for I never can get one thing to stick permanently to another. So back we went to G.Y.S., wondering how many weeks would elapse before they could lift us out again, for their facilities were much in demand and generally were booked up for some time in advance. However, Bob Petersen, the manager and a director, seemed to be surprised and distressed about the glue failure, and immediately had us lifted out again, but on the smaller, though equally efficient, screw-lift. This time metal strips were used instead of wood to hold the spoilers from slipping off, and they were welded to the rudder. So quickly was the job done, including the making of a new pair of spoilers, that we were lifted out and lowered back in again the same day, for all were alert to the harm that might be done to our new antifouling paint if it was left for too long exposed to the sunlight.

On sailing trials the spoilers seemed to work as well as could be expected, though they must have reduced our speed, and although the ship still carried a lot of weather helm on some points of sailing, when running she no longer showed any particular desire to turn to port. We therefore felt we should press on for Panama without much further delay. But first Susan made several trips by outboard over to the town of St. George's to buy as many stores as she reckoned she could stow away (Plate 8), for this would be our last sterling area for a long time, and once we had left it everything we bought would have to be paid for in hard currency. She found St. George's a good shopping area and handy for the dinghy, as this could be left just across the road in front of the shops, so long as she could find anything except parked cars to which to tie its painter.

I think Bob Petersen and his co-directors of G.Y.S. must have realized something of the worry and expense to which our new ship was putting us, for when we went to the office to pay our bill we found that the firm had been more than generous in its interpretation of what we owed.

Our last two nights at Grenada were spent on the island's south coast in peaceful L'Anse aux Epines (Prickly Bay) off John Slominski's attractive and hospitable home. But while we lay at anchor there a new plastic jerrican, which we had just bought, filled with five gallons of paraffin, and stowed in one of the cockpit lockers, sprang a leak and silently let its contents seep through the sound-proofing tiles down into the engine-room below. By now I had grown accustomed to mopping out that bilge, and found a sponge an admirable tool.

On two earlier crossings of the Caribbean on the way to Panama we had met bad weather with winds so strong that we ran before them under bare poles. Both trips had been made in early January, when the trade wind is usually at its strongest; but this time we did not leave Grenada until the third week of February, and only once did the wind exceed force 4, and often it was less than that. Because of the improvement to the steering, the helmsman no longer had such an awkward job, and even when running could often leave the wheel for a little while without a gybe as the result. We took four days to make good the 400 miles to Bonaire, one of the Netherlands Antilles which we had not previously visited, and approached it from the north-east, on which side its vegetation consists largely of cactus. Off the town of Kralendijk the chart showed water so deep and the bottom sloping so steeply that it seemed impossible to anchor, but we had heard that 'marina facilities' were now available. I fancy such rumours tend to get exaggerated, for all we could find was a vacant mooring buoy off the beach-club south of the town. However, that was just what we needed, and a Canadian who was working for the club kindly came to help us secure to it. He was skippering a remarkable little launch which had a wooden canopy larger than itself, and was taking a party of guests for a trip. A slight swell was running in the harbour, and somehow the launch's canopy got wedged under our massive bowsprit; for a moment it seemed that a capsize was inevitable, but the skipper quickly persuaded his passengers to move to the other side to heel the launch and so got free. We landed at the club pier, and the manager, who had lost his yacht there some years before, offered us the facilities of the place gratis, but we did not like to take advantage of this. At dusk a searchlight was trained on us, presumably for the entertainment of the guests, so for a change we had our evening drinks below instead of in the cockpit. The port officials seemed to wish us to leave as soon as possible, and as we found their barren-looking island and its tumble-down little town, where we did buy some excellent bread, rather dreary, we cleared out after only one day there.

By late afternoon we were in the lee of Curaçao in company with a small cutter, which Susan recognized as one that carries fruit and vegetables from Venezuala to the floating market at Willemstad. She was under jib, engine and awning, and had the legs of us, for there was not much wind. A thunderstorm was brewing, and seeing no reason to spend an uneasy night at sea when we might be at anchor, we put into Spaansche Haven, an excellent natural harbour with a twisting, narrow entrance channel, on the south-west coast of the island. Having entered the Dutch islands at Bonaire and had our passports stamped there, we supposed no further formalities would be necessary here. But next morning, after a bump which shook the ship and knocked some of the paint off the side, we were boarded by an irate official who said that yachts were obliged to enter and clear at each and every island, that Spaansche Haven was not a port of entry (we have since learnt that it is) and that by failing to go to Willemstad (the oil-impregnated capital) we had broken the law. However, he was just as concerned as had been the officials at Bonaire to get rid of us as quickly as possible, so we left that afternoon without going ashore.

We passed by Aruba, third and last of the Dutch islands, where we could see the glow of the oil refinery and the lights of tankers coming and going, and by night rounded the unlit Monges Rocks to run parallel with the South American coast, and set one of the twin running sails to help us along, but soon had to take it in again as the wind increased to 25 knots. We had hoped to stop at Cartagena in Colombia, but as we drew near to the longitude of that city the only bad weather we experienced during the passage overtook us, and in gale conditions with poor visibility and a heavy sea we did not care to close with the coast; instead we held on direct for Cristobal at the north end of the Panama Canal, running under the reefed mainsail only. Because we no longer suffered from the dangerous steering bias to port, steering in these conditions was not so difficult or dangerous as it had been before the spoilers were fitted to the rudder; nevertheless the helmsman's job (we felt this might be rather too much for the Pinta) was not easy and called for concentration. I was becoming more and more convinced, in spite of the improvement we had made, that a semi-balanced rudder and huge propeller aperture (which I had mistrusted from the start) were wrong for a sailing vessel, and that one day we would have to chop the balance part off the rudder and fill in much of the aperture. But we might have to put up with things as they were for a long time yet, as it was unlikely that we would be able to afford to have such work done in the U.S.A.

After the gale had blown itself out we had very little wind, and the 700-mile trip to Cristobal took us 6½ days. We arrived as one in a procession of ships shortly before midnight, and as we passed in between the two breakwaters which shelter the huge harbour, with a tanker and a freighter breathing down our neck, a launch rather longer than *Wanderer* ranged up alongside, a searchlight blinded us and an American voice asked:

'Have you been here before, captain?'

'Yes,' I replied, 'but not in this ship.'

'Okay. Then you can go over to the Flats and anchor.'

The well-handled launch then came within arm's length and

a fist-full of papers were handed to me, for Susan was steering, as I always like her to when entering port.

'Get these filled out and I'll be over before you go to bed.'

We motored for three miles across the harbour to its eastern side, and let the anchor go in 4 fathoms south of the docks and inshore of the fairway buoys. After we had switched the engine off the silence was intense; no breeze stirred, no sound came from the shore; the air was heavy and redolent of the surrounding jungle as we stowed the sails and coiled down.

We had only just finished the paper-work (six copies of some forms were required, but carbon paper had thoughtfully been provided) when the launch arrived and the boarding officer stepped aboard. He was efficient, but helpful and understanding, especially when it came to a de-rat exemption certificate, which we had not got, and a de-bug bomb, which we did not want used in our clean and polished accommodation. He had a cup of coffee and left us at 0130 with the reminder:

'The admeasurer will board you at six o'clock.'

We got a few hours of sleep, and had just cleared away the breakfast things when the admeasurer came, a gentle, soft-spoken man who with his wife had lived in the zone for twenty years and loved it, but soon was due for retirement. With a tape measure and a little help from me he took *Wanderer*'s vital statistics, for although canal tonnage is much akin to British register tonnage, it is not identical, and every newcomer must have her measurements taken so that a tonnage certificate, on which canal dues are levied, may be issued.

Even when the business of entry has been completed, the yacht calling at Cristobal may not go to the Panama Canal Yacht Club, which provides the only comfortable and convenient berth, until permission has been obtained at the custom-house, and the odd thing is that although the captain (as I had been called by the boarding officer) is considered sufficiently proficient to pilot his yacht through the traffic and across the harbour by night, he is not permitted to use an outboard motor on his dinghy unless he has a pilot's licence; he must row; this takes half an hour and may be impossible if the wind is strong.

Mrs. Nott, assistant secretary of the club, who with her husband had arrived by yacht in the Zone five years before, was most helpful; although it was Sunday she opened her office to get us our mail, and showed us a berth which *Wanderer* was soon occupying with her bowsprit almost poking into the well-remembered club restaurant, where we found the 'sizzling' steaks just as superb as ever they were, and the staff just as cheerful. There, where the juke-box blares all day and through most of the hours of darkness—many employees of the Canal Company are members, and there is usually one or another who wants to relax with a Coke and a little light entertainment as he comes off day- or night-shift—we remained for three days.

We had already been lent some of the charts of the west coast of North America, and intended to buy the remainder at the local chart depot, but Alvin Daniels, who had recently come down the coast with Carleton Mitchell in the latter's motor cruiser *Sans Terre*, kindly gave us all that we needed together with the appropriate volume of sailing directions, and he went through them with us, marking anchorages and making useful notes. I had arranged for a copy of the current *Nautical Almanac* to be sent to me at Grenada, but it never arrived; so throughout the trip to Panama I had been navigating with the previous year's almanac. Instructions to cover this contingency are given in the book and are not difficult to follow, but they add a little to the calculations, and do not permit the book to be used for moon or planet sights. I was therefore glad to be able to get a copy of the up-to-date almanac in the Zone.

At the club we met Walter Reinheimer, a senior pilot who wanted very much, he said, to take us through the canal; he asked for the job, but did not get it, for clearly he was considered too important a man for such a small vessel. We visited the dispatcher's office and arranged for a centre lockage, and paid our dues, which came to $15.00, and that was just what it cost to get one of my teeth re-stopped in the Zone. Most small-craft people regard a transit of the canal as a slightly hazardous undertaking, for the pilot is in absolute charge, yet he may not be familiar with the handling of small vessels, especially those with limited power. But on our two previous transits we had been assured that the yacht was fully covered by the company for any damage that might be done to her. I was therefore surprised on this occasion to have to sign an indemnity form which said in effect: 'If we smash you up you pay the bill.' For a centre lockage a hand to tend each of the four lines, plus one to steer, is obligatory. Fortunately *Bluebird of Thorne* was with us at the club, and her owner, Lord Riverdale, kindly let us have three members of his crew to make up the required number.

Our pilot, who was comparatively new to the job, came aboard together with our extra crew at 0600. He asked if his wife might join us at Gatun for the trip, and of course we said she would be welcome; he also asked us to keep the awning spread, and although that would certainly make life more comfortable in the cockpit, it would rule out any possibility of sailing across the lake, which we had always done before, and which most pilots enjoy. We mounted the three locks at Gatun (Plate 9, *top*) while my heart beat rather faster than usual, for it is here during the inrush of water that there is the greatest risk of a small vessel getting into trouble. But all went well, and having risen 85 feet above the sea, we went alongside a pier and disembarked two members of our volunteer crew, for their help would not be needed in the down-locks where the going is easy, and as *Bluebird* was due to make the transit next day, I felt it imperative to return her people to her as soon as possible; but we asked the third member, Michael Hart, to remain with us as we knew we would be glad of his efficient help, and he seemed quite happy to come. The pilot's wife then boarded us, and we set off.

Shortly after starting the 28-mile crossing of Gatun Lake we passed, lying at anchor, the huge bulk-carrier *Mythic*, a Monrovian ship of 62,000 tons, the largest the canal could accept. It turned out that our acquaintance Walter Reinheimer was in charge of her that day. She soon weighed, and as she overtook us, with the steering tug at her stern looking like a midget, Walter sent us a good luck message over our pilot's walkie-talkie radio which, standing on a cockpit seat, was kept tuned in and switched on throughout the day. But in Gaillard Cut, the narrowest part of the canal at the south end of the lake, where earth-moving machinery was at work coping with the usual landslide there, our pilot decided to overtake the *Mythic*. Due to the great length of that ship and to the very slight difference between her speed and ours, the operation took half an hour

(Plate 9, *bottom*). Susan and I felt very anxious, and so I think did someone else, for at one point a voice over the radio said:

'With so big a ship that is a foolish thing to do.'

It was also a pointless thing to do, for on arrival at Pedro Miguel (Peter McGill, as the Americans call it, first of the three down-locks) we had to tie up to the wall and wait for *Mythic* to come past, and we felt rather vulnerable, as there was not much room for the tugs to manoeuvre in as they nudged the vast ship into the lock where there was only a nine inch space between her plates and the concrete walls, through which something more than 62,000 tons of water had to escape. At the height of this fine display of seamanship Walter found time to hail us from his lofty bridge to say that he had taken some photographs of us, and where was he to send the prints.

We shared the down-locks with a Japanese ship and a tug, alongside which we lay as the water gently subsided, and at dusk, having taken 11½ hours for the 40-mile transit, passed under the great span of the Bridge of the Americas, and secured to one of the guest moorings off the Balboa Yacht Club. Our pilot and his lady left us, and after dinner we sent Michael by train back to *Bluebird* with considerable regret, for he was such a keen, efficient and likeable young man that we would much have enjoyed his company and help on the next, and rather difficult, stage of our voyage.

# 3

# Coastwise in Central America

As the prevailing wind along the Pacific coast of North America is north-west, a sailing vessel bound from Panama towards San Francisco, as we were, can make the coastal passage only with great difficulty; so the proper thing for her to do is to stand out to the westward of the Galapagos Islands, about 1,000 miles away, before turning north, and then not to steer direct for her destination until she is on or near its latitude. But that is a long and tedious passage without much interest, and as *Wanderer* had a reliable diesel engine and a plentiful supply of fuel for it, and as Susan and I wished to see something of Mexico, we had decided to keep to the coast at least as far as Acapulco, for according to the U.S. pilot charts we should have the current with us up to that port; we might even reach Zihuatanejo. But we agreed that if thereafter we found the struggle against both wind and current too much, we would head out to sea until we reached the sailing route from the Horn to San Francisco, and follow that.

We had left our fascinating berth at the south end of the Panama Canal close beside the deep channel through which a constant stream of ships of all sizes, shapes, colours and nationalities made its purposeful and almost silent way by day and night, on 18 March, and went out 10 miles to the Island of Taboga,

▶

9. *Top*: Entering the first of the locks at Gatun on the Panama Canal astern of the ship with which we were to share it. It is in the up-locks that a small vessel is most likely to get into trouble because of the inrush of water as they are filled. *Bottom*: 'With so large a ship that is a foolish thing to do.' In Gatun Lake our pilot decided to overtake the 62,000-ton bulk-carrier *Mythic*, the biggest ship the canal could accept.

YTHIC

where we launched the dinghy and cleaned the oil of Balboa off the sides. The next day a fair wind carried us out of the Gulf of Panama, and as by night we rounded Punta Mala, the gulf's western point, we entered for the first time this voyage waters which were entirely new to us.

'I feel excited,' said Susan, as she took a warm hand off the wheel and touched my shoulder. 'I feel the real voyage is only just beginning.'

I could not see the expression on her face, for the night was dark and the luminous grid compass by which she was steering shed no light.

'I make it about 3,500 miles along the coast from here to San Francisco,' I said as I squeezed her hand, 'and in all that distance we do not know a single soul.'

'Well,' she replied cheerfully, 'if the Americans are anything like as nice as those we met along the east coast in little *Wanderer* they'll be pretty good.'

Dawn revealed a chain of 3,000 to 4,000-foot mountains not very far from the coast on our starboard side, but this soon vanished in mist. In general we found that visibility was poor, and this was aggravated by the fact that it was the dry season in which forest fires raged, large areas of vegetation were burnt off in preparation for the new crops, and for many hundreds of miles the land was enveloped in smoke through which the sun shone dimly like a flourescent orange. Usually, as on this first day, only at dawn and dusk did the mountains show themselves, grey, aloof, mysterious, (Plate 10) and sometimes on the off-shore breeze the scent of burning wood drifted out to us. We experienced this indifferent visibility all along the coast of Central America and up the Mexican coast until we reached the desert land of Baja California, and at times it added to the difficulties of pilotage, for the mountains are valuable and sometimes the only recognizable landmarks.

Alvin Daniels, Tom Steele, and other Americans familiar with these waters had marked our charts—a mixture of British and

◀

10. Because of the persistent mist along the coast of Central America, it was usually only at dawn and dusk that we could see the mountains, grey, aloof, mysterious; the sun shone dimly like a fluorescent orange.

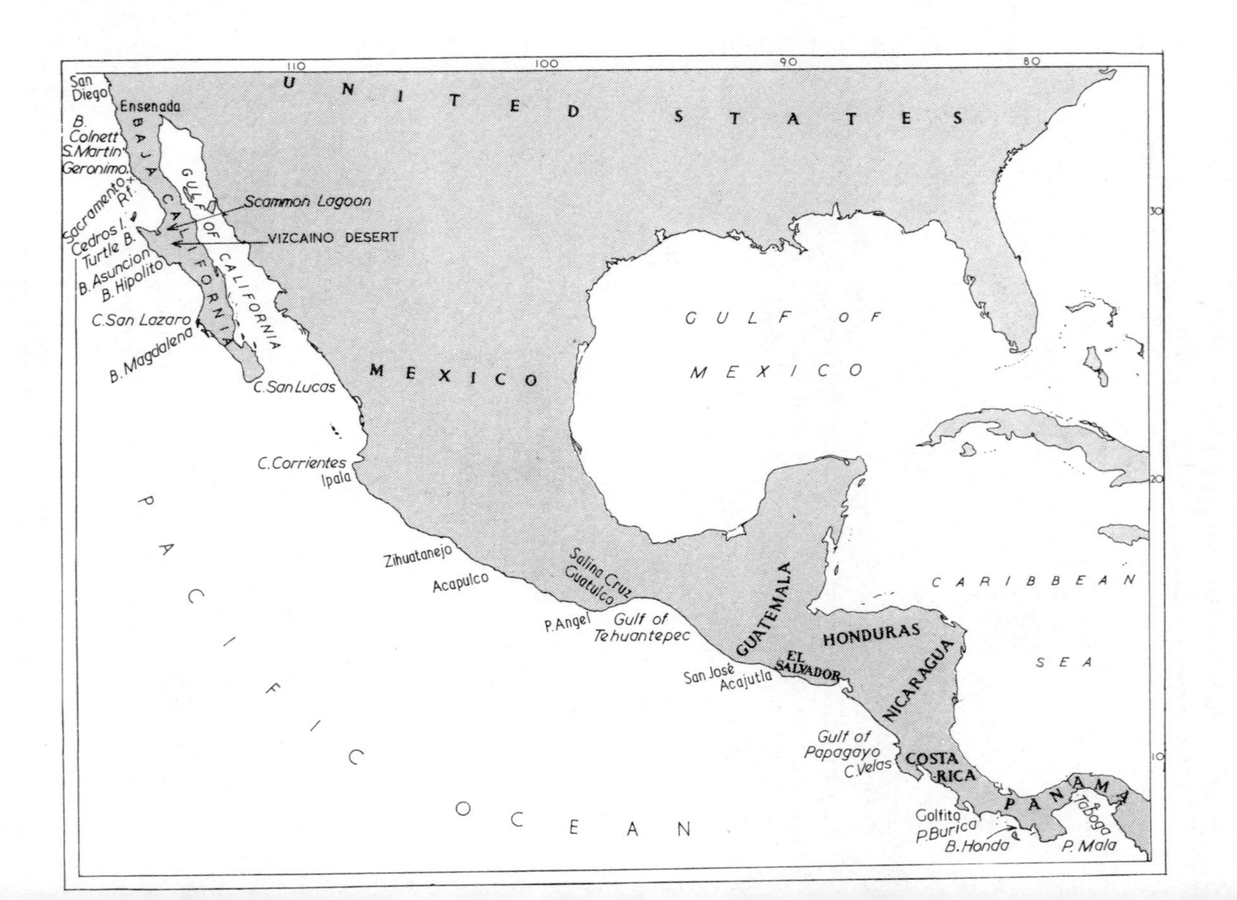

110
100
90
80
30
20
10
UNITED STATES
San Diego
Ensenada
B. Colnett
S. Martin
Geronimo
Sacramento Rf.
Cedros I.
Turtle B.
B. Asuncion
B. Hipolito
C. San Lazaro
B. Magdalena
C. San Lucas
BAJA CALIFORNIA
GULF OF CALIFORNIA
Scammon Lagoon
VIZCAINO DESERT
MEXICO
GULF OF MEXICO
C. Corrientes
Ipala
Zihuatanejo
Acapulco
Salina Cruz
Guatulco
P. Angel
Gulf of Tehuantepec
San José
Acajutla
GUATEMALA
EL SALVADOR
HONDURAS
NICARAGUA
Gulf of Papagayo
C. Velas
COSTA RICA
PANAMA
Golfito
P. Burica
B. Honda
Taboga
P. Mala
CARIBBEAN SEA
PACIFIC OCEAN

U.S.—with suggested anchorages, and that evening we made use of the first of these, an open, unnamed bay 155 miles from Taboga, where we brought up in 7 fathoms. Tumbling hills backed the black sand beach, which the swell laced with white. Several sharks, a turtle, and three yellow and brown sea-snakes came to investigate us, and at dusk the birds in the dense forest whistled like a lot of kettles. We were, as the weeks went by, to grow accustomed to such open anchorages, often devoid of any sign of human activity, and always with the roar of surf in our ears. But at that time of year, between the end of the cyclone season and the beginning of the rains, anchorages are regarded as safe so long as they are sheltered from winds between north-west and north-east.

Our anchorage for the next night was in marked contrast, for, having sailed 50 miles, we entered one of the very few perfectly sheltered harbours on the coast, landlocked Bahia Honda. There we found an anchorage behind the island which lies in the middle of it and on which most of the fishermen and their families live to avoid, it is said, the snakes and other dangerous creatures of the mainland. Almost at once we were surrounded by canoes carrying quiet, handsome Indians, who seemed to want little more than just to sit and stare at very close quarters. No doubt a yacht is not a common sight there. One man, however, did ask for cigarettes, and when I made him a present of a packet he wanted to know how much he should pay me for it. Susan had some sweets for the numerous children, and the people in one canoe gave her a fish and some bananas; brown hands offered brown eggs, but she would not accept these as we already had more than they. Our visitors did not leave until nightfall; when they had gone utter silence prevailed with not even the whisper of a ripple on the beach to disturb it.

That day, while sailing to Honda, we had noticed marked on the chart 30 miles south of our course a magnetic anomaly; the sailing directions informed us that this increases the normal 5° easterly variation by 13°. But our compass was swung in the opposite direction, and showed a westerly variation of about 10°. We guessed this must be due to our being on the northern instead of the southern side of whatever it is that causes the magnetic disturbance. However, some of our landfalls had not been so good as they might have been, and we were beginning to think that our compass had more deviation than the adjuster had tabulated, and I thought we should check it with an amplitude —a bearing of the sun at rising or setting—but due to the haziness of the sky an opportunity for this never presented itself. We did manage to check it roughly with transits the day we left Honda, and it appeared to be normal, with one group of rocks after another appearing in the right place. Nevertheless the 140-mile passage to Golfito—a banana-loading port at the head of the Gulf of Dulce—proved awkward. According to my reckoning the 12-mile light on Punta Burica, where we needed to alter course for the gulf, was abeam 4 miles away at 0230; but although the night was dark and visibility seemed to be no poorer than usual, we did not sight it. This lack of confirmation of our position of course raised doubts in our minds; could there be another unmentioned magnetic anomaly affecting our compass, or was a vagrant current responsible? We did not care to alter course inshore until dawn, and when daylight came there was no sign of land. Four hours later through the mist we could just make out ahead the line of the coast fringed with heavy surf, and in front of it a small black rock, which we understood to lie off the gulf's western entrance point; we did not see the eastern point though it was only 5 miles away.

We motored for 15 miles up the gulf in glassy calm, sighting some sea-snakes, and turtles lazily sculling themselves along on the surface; there were sharks—these are said to make any attempt to cross the bar of the near-by Coto River by boat dangerous—whales and schools of swift porpoises; the sea was full of life. The heat was intense; the saloon thermometer registered 104°F., and our bare feet could not stand on the teak deck except in the meagre patches of shade.

Costa Rico's Golfito, just within the border between that country and the Republic of Panama, is a large, natural, landlocked harbour with the tree-clad mountains rising steeply from its shores. At that time three ships a week were putting in to load (mostly) bananas at the modern installation which was served by railways leading from the plantations. The office of the captain of the port was in a little café two miles from the port; there we went to enter, and a second visit was necessary to obtain the obligatory clearance; but neither cost us anything except time.

We also called on 'Captain Tom' whose story we had learnt from another yacht. This kindly, one-legged American used to do towage work up and down the coast with an ex-submarine-chaser which he owned. Wishing to settle in gentle Costa Rica, where taxation was low and far more money was spent on education than on defence, he obtained permission to put his ship ashore while he dealt with some defect. He did the job thoroughly, and at full speed ran her up the beach under the trees at the west side of the harbour. That was fifteen years ago, and now she lay a rotting hulk, while many of her fittings had gone into the house he built with other parts of her. His home was also the school, and in his spare time he built ferro-cement boats.

When we told him we had not seen the flashing light on Punta Burica he laughed.

'That's been out for at least a year,' he told us, 'and as you go north you will find many others like it.'

He was quite right, and I think that at least half of the lights for which we searched on the coasts of Central America and Mexico were extinguished, while many others were of such low power that they were visible for only short distances. American charts disregard candle-power and base the range of a light solely on its height; as many of the lights were high, they were credited with a visibility in excess of 20 miles, yet their paraffin-burning lamps were of such low power that their range might be no more than 2 or 3 miles. Indeed, on several occasions we noticed that the light exhibited was of the same power as that showing from the window of the light-keeper's dwelling. We soon learnt not to rely on any of them, and to navigate by night with the greatest discretion.

On leaving Golfito our intention was to cruise along the western part of Costa Rica, where north of Cabo Vela some pleasant anchorages are to be found. On the way towards it, and during the night, we became involved with a violent electric storm which lasted for six hours. At its height each blinding flash of lightning was followed so closely by its crash of thunder —indeed, the two were sometimes simultaneous—that we feared we might get struck, though I suppose a steel vessel with steel rigging should be reasonably safe. The storm was accompanied by torrential rain, and twice the wind, which fortunately never blew very hard, boxed the compass, leaving us with a sense of bewilderment as we tacked and gybed and the glass of the compass steamed up. After it had left our neighbourhood the storm continued to rumble and grumble over the land for a long time, and it did nothing towards clearing the air or relieving the oppressive humidity. It left behind a veiled sky, and the sun had a sinister halo.

Later, as we approached Cabo Vela, a strong wind sprang at us from out of the north-east and rapidly freshened to 43 knots. This was a Papagayo, a local gale named after the gulf that breeds it, and in which we now were sailing. At the time of its onset evening was approaching, and although an anchorage (of sorts) lay only 10 miles away, we could not reach it before dark as it was to windward of us. We therefore hove-to for the night, trying various combinations of sail, and finding that we lay best under the reefed mainsail only, but then with the wind only just forward of the beam instead of 45° as it should be. I believe *Wanderer* would probably heave-to better if the mainsail were taken in and the mizzen and staysail set instead, but that would involve a lot of labour at a time when one is least inclined to undertake it. By dawn we had been blown away out of sight of land, and the day after that, being by then even farther from the shore along which we had hoped to cruise, we abandoned the idea, took the mainsail in, set the staysail, and with the wheel lashed ran on towards the Gulf of Tehuantepec. This large body of water, assisted by the high pressure area in the Gulf of Mexico, breeds winds which are known as Tehuantepecers and sometimes reach hurricane strength. But by now the end of March was near, and over a period of 12 years only 16 Tehuantepecers of from force 8 to force 12 had been recorded in the month of April, so we felt the chance of encountering one was small. However, we learnt later that a Tehuantepecer was blowing at the same time as our Papagayo, the two combining to blow over a very wide area. One yacht reported being in it 500 miles from the coast, and another, which was close to us at the time, the trimaran in which Michael Kane was nearing the end of a world voyage, was damaged.

Some 300 miles on, the gale having abated by then, and we having passed Nicaragua (which is said to have pirates and to be unfriendly to yachts) and Honduras (which has only a few miles of Pacific seaboard), we closed with the coast of El Salvador near the point of Acajulta. Alongside this place on the chart someone had pencilled a note: 'New yacht club asks yachts to stop.' The roadstead is open to the ocean and to the prevailing wind, and a considerable swell was rolling in to break on the beach. Two freighters lay at anchor off the iron pier with lighters in attendance. If in fact a yacht club was there it seemed impossible that it could have any yachts or berths for them; later we were told that there was a club, but that it was situated on a lake some distance inland. We continued on our way, sailing fast closehauled during the afternoon onshore wind, and motoring through the calm of the night. In the dark we passed San José on the coast of Guatemala, another open roadstead where ships were anchored off and the swell was running high, and next day came into Mexican waters, having passed by four countries in as many days.

There we entered the Gulf of Tehuantepec, the shore of which sweeps round in a great curve, and at one point is only 120 miles from the Gulf of Mexico. For 300 miles it consists of a grey sand beach, which in all that distance is unbroken by a single rock or headland. The beach is backed by a wide strip of low country in which are many lagoons, but the entrances to these are blocked by sand bars in the dry season, and these bars often project some distance seaward. Beyond the low land rise the mountains except in one place, where a survey was made for a possible ship canal through to the Gulf of Mexico.

We had been advised to coast round the gulf and keep in 4 or 5 fathoms, so as to avoid the contrary current and to be close to a weather shore in the event of a Tehuantepecer blowing. For two and a half days, mostly motoring in calm, we made our way round the gulf, but failed to keep as close to the shore as we had been urged to do because of the nearness of the breakers and the danger spots where bars projected seaward, and for the most part we stayed on the 10-fathom line which, particularly during the hours of darkness, seemed to us quite close enough. It was a strange and sometimes beautiful journey.

At twilight, when the mountains showed faintly through the haze, great flights of silent pelicans crossed our course inward- or outward-bound, lifting and dipping in stately unison. Occasionally one or more of these large birds let down its flaps and alighted with a splash which looked like a crash landing, to pack a fish into the pouch of its lower bill and then take off again with one single, powerful downward beat of wings which seemed to spring it into the air. Several shrimpers with pelicans in attendance passed us, and once we saw one at anchor very close to the breakers; her boat was afloat, and on the beach were people and horses, and some sort of market appeared to be operating; so landing through the surf must have been possible at that particular spot. On another occasion we saw on the sand bar which closed one of the lagoon entrances a knot of figures busy with a net. Otherwise there was little sign of man.

Eventually we reached the end of that long, long beach, and passing headlands which washed their feet in the sea, came to the artificial harbour of Salina Cruz and anchored in its outer part. We had been advised not to enter Mexico there because of the high charge made by the officials, so we did not fly international code flag 'Q', nobody bothered us, and after a good sleep, which for once was not disturbed by the roar of surf which had been in our ears for so much of that passage, went on again with a strong wind and an unexpected fair current to Guatulco, which we easily identified on account of the little lighthouse which stands near its entrance. Susan and I belong to the Seven Seas Cruising Association, an American organization which has as members only those who live aboard their vessels and cruise; all are given the rank Commodore so that they cannot justifiably complain at the way the thing is run. In its monthly *Bulletin* we had read many accounts written by Americans of cruises in Mexican waters; but none of them mentioned this pleasant little cove, possibly because no American chart of it is published; however, among our British charts we found a large-scale plan. The cove is about half a mile long by quarter of a mile wide, and as it is open only to the south-east it is not exposed to the westerly swell which is the curse of this coast. A sand beach lies at its head, and there in the midst of a straggling line of fishermen's shacks stood a blue shrine shaded by palms; the fronds of these provide the only touch of green in the dead-looking landscape of the dry season, when grey and brown are the predominant colours. Almost to the day 390 years before our visit, Drake with the *Golden Hind* during his circumnavigation put in at Guatulco, sacked the town, pillaged the church, and took aboard sufficient water for 50 days.

Once again we had a fair current when we left, but this was the last we were to enjoy for more than a year. We went to Puerto Angel, a cove smaller and less well protected than Guatulco, where the swell broke angrily on the rocky shore which was too close for our peace of mind. There we officially entered Mexico, and the business of entering took two and a half hours, involved the labour of six men with six typewriters who made six copies of everything, and cost the regulation $16.00; this has now been increased to $40.00. No doubt if we had possessed a working knowledge of the language the business of entry might have been speeded up a bit, but naturally enough the civil servants at Puerto Angel are in no great hurry since Salina Cruz has taken all the commercial trade, and the only papers now to go into their huge, dusty files are those from a few yachts.

The 216-mile passage to Acapulco took us 51 hours of almost continuous motoring against a light headwind and a tiresome jump of sea. I passed the time reading in *Harper's Magazine* Norman Mailer's grisly *Miami and Chicago*, and was shocked at U.S. police methods, particularly their use of jeeps shod with barbed wire for controlling mobs, and I did begin to wonder what sort of a country we would find after Mexico had been left astern. Although none of the five intermediate coast lights were working, and although, perhaps because of deviation of the compass, we got rather too far offshore, we had no difficulty in finding Acapulco as the loom of that town's lights could be seen in the night sky some way off. Soon after breakfast on the second day we entered the big, circular harbour, and anchored near the yacht club off some high-rise blocks of apartment buildings which, I believe, occupied the place where the Manila ship (or galleon) used to berth while loading and discharging her cargoes. We did not make use of the club because of the high cost, but, like the people from the other visiting yachts, landed on the near-by beach and made our way to the street through a friendly little shipyard, where one or other of the smiling shipwrights always came running to help us if we had much to carry.

It was fun to take one of the cheap, frequent, and mostly unsilenced buses into the tourist-crowded town to do our shopping and other business—the contrast was remarkable after the lonely, empty coast we had been following for so long—and to get to know the people in the other yachts. Among these were Walter and Katie Maertins, an American couple, in the 46-foot ketch *Evening Star*, which they had built with their own hands twenty years before. They had just come in direct from the Galapagos Islands and were homeward bound for San Diego in California.

On our voyages we have made many friends, but always among them are a few very special ones, such as the Greys of *Altair* and the Kittredges of *Svea*. The Maertins were to join this little group of special people who think the same way as we do, and with whom we can laugh, and if necessary cry, at the same things. They knew the Mexican coast well, for this was their fourteenth cruise along it. We told them that we intended to coast north as far as Zihuatanejo to see the place where in 1741 Anson had watered, wooded, and provisioned his ships before making the Pacific crossing to China, but that if progress became too difficult after that we would probably put to sea and stay there until we reached California. They laughed at such a crazy notion, and said that when they went north they kept 'one foot on the beach' and spent nearly every night at anchor. If we remained in company with them we could do the same, and they would be glad to have us along. Over a drink we agreed to do that provided we could endure the hard headwind, but at the same time we felt we had to make the point that *Wanderer* was supposed to be a sailing and not a power vessel—we changed our minds about this a little later.

Before leaving Acapulco we wanted to top up our tank with diesel fuel. Although the motorist visiting Mexico may buy fuel when and where he wishes, the yacht must first obtain a permit, so on one of her shopping expeditions Susan called at the customhouse to get application forms. These were not available, but she managed to borrow an old used one and brought it back on board to make the required six copies on the typewriter. When she returned to the customhouse I went along with her. The fat young woman whose job it was to stamp the application forms kept us waiting for forty minutes while, seated at her desk, she slowly unwrapped and consumed the contents of a box of chocolates. Before the permit could be issued we had to pay 102 pesos (a little over £3), and when we asked why this should be the curt reply was 'Overtime'. It seemed that only at Acapulco was a charge made for a fuel permit.

The fuel pumps were situated on the quay about a mile from our anchorage, and Walt and Katie and the Pecks from *Kaisun* kindly came with us when we went to take on fuel, for they knew how difficult the operation would be and that we could not possibly manage on our own. One might have supposed that the quay abreast the fuel pumps would be kept clear of other craft so that vessels requiring fuel could get alongside it. But—and this was Mexico—a cluster of small launches lay there on moorings with their sterns secured to the quay, and to get within reach of the pump we had to anchor and manoeuvre in stern-first forcing a passage for ourselves. Our friends had brought their own boathooks to fend off our sharp-gunwaled neighbours, and Walt had thoughtfully provided a length of hose as he guessed, correctly, that the hose of the pump would not be long enough. Although Acapulco is a fine harbour it is not land-locked, and that day there was quite a surge in it; this was particularly noticeable at the quay, where it put a considerable strain on our necessarily short stern-lines. Meanwhile high-speed launches towing bronzed water-skiers buzzed us, their steep wash grinding the moored boats harshly against one another, and the air vibrated with the roar of their exhausts. But thanks to our friends we obtained the fuel without getting damaged, and returned to the anchorage to prepare for departure, and to watch after dark through binoculars a remarkable, floodlit exhibition of pole climbing on the top of a distant hotel. In another direction away from the town and high up on an unlit hill the red and white lights of a radio tower glowed like jewels on black velvet.

One of our most prized possessions on board is a copy of the second edition of *A Voyage Round the World* by George Anson Esq., published in 1748, and given to us by a generous antique dealer in Barbados some years ago. It contains the original theme of many a good boys' story, and accounts of some very remarkable feats of seamanship and endurance. One of these concerns the *Centurion*'s cutter, an open boat about 22 feet in length, manned by a lieutenant and six seamen. She was dispatched from Chequetan to watch for the departure of the Manila ship from Acapulco 110 miles to the east. In the event of the ship leaving port the cutter was to return at once with the news to Chequetan, whence Anson reckoned his ships with their superior speed would be able to overhaul and take her. The cutter was ordered to cruise out of sight from the shore off Acapulco for 24 days, and if she had not seen the Manila ship by the end of that time was to rejoin the squadron. She carried out the first part of this order, but then an adverse current forced her away to the eastward. Short of both food and water, her crew repeatedly closed with the inhospitable shore but failed to find a possible landing place as the breakers were too heavy everywhere. The blood of turtles and a providential downpour of rain (a most unlikely happening in the dry season) saved their lives, but they were weak and ill when eventually they were found by the squadron, for they had been at sea for six weeks. Knowning as I now do that part of the Mexican coast at the same time of year, when by day the sun beats down relentlessly, the exploit seems even more remarkable, and I often thought of the *Centurion*'s cutter as we forced our way to the westward against a fresh headwind and a steep sea, bound from Acapulco to Chequetan in company with *Evening Star*. I am sure we could easily have found and entered that place (it is now called Zihuatanejo, which is pronounced something like Say-what-an-hero) with the help of Anson's description alone, so detailed and accurate is it; but we did have a large-scale chart. The lagoon at the harbour's northern side, from the inland end of which Anson was able to take water sufficiently fresh for filling his ships' water-casks before sailing for China, we found to be completely dry, but a footbridge at its seaward end suggested that it did still fill up at times, probably only in the rainy season; the locals, however, were getting their water, largely on donkeys, from a near-by well which probably taps the same spring as served Anson. The high, sand beach separating the lagoon from the harbour, across which the *Centurion*'s crew rolled their casks to the waiting boats, was still much the same as it must have been in their day; even the dugout canoes remained unchanged, though most of them were fitted with outboard motors (Plate 11).

In the harbour fish were so plentiful that even we, who are no great fishermen, were able to catch some for Nicholson simply by dangling an unbaited hook over the side. But the local fisherman used a more efficient method. He had a circular net a fathom in diameter with small weights along its circumference, and this he dropped horizontally into the water from the end of the town pier. The net just fell on the fish, whose gills got caught in the fine mesh, and was immediately hauled up with, perhaps, 50 fish entangled in it.

The little town was something of a tourist attraction, with big, grey motor-coaches rumbling in from Acapulco to stir up the dust, and in the evening it was defiled by the brazen, amplified voice of a man at a fun-fair against which the church bell could scarcely make itself heard. We visited *Evening Star*, and as the Maertins were fond of cats we took Nicholson along. He cased the joint, accepted a handout with appreciation, and, deciding that *Evening Star* and her people were okay, settled down on Katie's bunk.

From Zihuatanejo we continued on our way along the coast. From this point on, and until we reached Turtle Bay 1,000 miles ahead, all but one of our anchorages were in open roadsteads; at each we lay in the lee of a headland which gave shelter from the prevailing north or north-west wind, but not always much protection from the swell. We sailed when we could, which was not often, and motored shamelessly when we couldn't. The wind, which always draws along the coast no matter whether this runs west, north-west, or north, usually springs up about noon and freshens until dusk; by midnight, or earlier, it dies away. Therefore when the distance between anchorages was 50 miles or less, we used to start as soon as there was light enough to see our way out of the anchorage (earlier if there were no dangers to avoid, for 'one hour in the morning is worth two in the afternoon', as we often reminded one another) in the hope of reaching the next anchorage before the headwind became strong. Occasionally the morning calm continued all day; but sometimes the wind got up early, and those were the bad days.

The two yachts usually left each anchorage together, and if there were dangers to avoid we were always glad to let *Evening Star* with her local knowledge lead the way, and it was often remarked by the Maertins that on those occasions *Wanderer* usually took a slightly wider sweep, and that she was rarely to be seen except out on their port quarter. Our engine ran very sweetly over most of its speed range but it did have a period of vibration between 1,500 rpm and 1,800 rpm. At the lower speed we slowly dropped behind, and at the higher drew ahead, so we and *Evening Star* frequently overtook one another.

Like most Americans Walt had a ship/shore radio aboard, and every day, whether we were within sight of one another or not, he called us at 0830 and 1630 to tell us how they were getting along, to notify us of any change of plan, or to say something about the weather (obtained by radio from fishing vessels) or the anchorage which lay ahead. We enjoyed this although we had no means of answering, and each day looked forward to the schedule. Sometimes Walt would say:

'This is fine and dandy. We're keeping our fingers and toes crossed and expect to get in around . . .' Or, 'Oh boy, oh boy, we sure are like a submarine.'

For much of the time the two yachts were within sight of one another, and it was agreed that if either wished to communicate with the other she would turn at right-angles to the course. The only time this was done it was *Evening Star* who turned, and as she had been having some trouble with her engine that day, we wondered if we might have to offer her a tow. So we altered course to close with her, and switched on our radio receiver; but nothing came over it except an unintelligible conversation between two Mexican fishermen, and after a few minutes our consort resumed her proper course. We enquired about this later, and learnt that the Maertins had spotted a Japanese glass fishnet float of such an unusual size that they just had to go and pick it up.

For a week we remained in company, exchanging at our anchorages visits, drinks and meals with one another, and learning more and more about Mexico and California from our knowledgeable friends. We had spent a night at Ipala, where we shared a huge hunk of venison, and were bound round Cabo Corrientes (Cape of Currents) for the next night's berth. This was one of the days when the wind sprang up early and soon was blowing hard, and as our course off the cape had to be altered for the first time in many days to a little east of north, naturally the wind hauled round too and so remained ahead. *Evening Star* manfully plunged on under power with decks awash, but Susan and I could not resist the temptation to put to sea and take this opportunity of a slant to cross the mouth of the Gulf of California (or Sea of Cortez), heading for Cabo San Lucas, 280 miles away at the southern tip of Baja California, which is the long Mexican finger that forms the western side of the gulf. What joy it was to ease sheets and feel the ship moving fast under sail, and to see the patent log spinning merrily. But we felt this could not possibly last for long, and sure enough as soon as we had drawn well away from the land the wind backed and headed us. Closehauled then we drove on uncomfortably, but we found the sea less steep than it had been inshore, and we made reasonable progress without resource to the engine.

The crossing took us three and a half days, and during the second day the rudder became so stiff that the motor of the auto-helmsman was overloaded and repeatedly popped its circuit-breaker; we therefore steered by hand, but I feared the strain of moving the rudder might cause damage to the steering gear, and felt I should do something about it. It may be recalled that we had experienced similar trouble in Spain due to the rusting of the upper part of the rudder stock inside its nylon bearing; but this time I found that the follower of the gland (the adjustable flange that holds the packing in position to keep the water out) being too tight a fit on the stock, had almost seized to it with rust. Two hours of contortions in the after-peak with spanners, freeing-oil, a hammer, and bleeding knuckles effected very little improvement, as I was unable to ease back the follower sufficiently to clean the rust from the stock, and we therefore felt anxious all the rest of the way to San Lucas. Although

*Evening Star* did spend a night in the chosen anchorage on the east side of the gulf, she made considerable use of her engine on the crossing and arrived only a few hours after us. Walt, who is an able engineer, kindly came to my help with heavier tools. With a lever under the follower he was able to lift the entire rudder assembly a fraction of an inch and hold it there, while I with his heavy, long-handled hammer drove the stock down with blows on its squared upper end. In this manner and little by little we forced the follower high enough to let us clean the rust from the stock with emery cloth. As a follower takes no load, it need not, indeed it should not, be a tight fit, and we were to have further trouble with it from time to time.

The San Lucas roadstead is better than many on that coast; a fantastic, jagged, high-key, windswept headland pierced by a great hole protects it from the west, and a long, curving beach of sand absorbs some of the swell—we have repeatedly noticed that anchorages where the shores are of rock are much less comfortable than those with shores of mud or sand. In the fine season it was a busy place; Mexican, and occasionally American, fishing vessels brought their catches of tuna to the fish-canning factory, and as fuel and water of good quality (this was rare in Mexico) could be had, it was a popular stop for American yachts bound to or from the Gulf of California. One of the yachts there during our visit was the 230-ton, two-masted schooner *Goodwill*, which in her day had twice been first to finish in what is known as the Transpac, a race from Los Angeles to Honolulu. We had seen her before at Acapulco, and now we noticed that her bob-stay had carried away at the bowsprit end; no doubt this did not matter much as the only canvas she appeared to possess was a fore-staysail, but as the bobstay had been left hanging from the dolphin-striker, it added to her generally run-down and neglected appearance. She was still there when we left in early May, and later we were to hear of the disaster that befell her. Ashore we found an air-strip, a hotel, and half an hour's walk inland along a sand track the little village where the port

▶

11. The beach at Zihuatanejo, across which Anson's men rolled their barrels of water, filled in the lagoon (now dry) beyond. The dugout canoes are much the same as they saw in 1741, but most of them now have outboard motors.

officials had their offices and where some stores could be obtained; at the post-office no stamps were available.

Between San Lucas and San Diego, which place 780 miles to the north-west would be our first California port, only three of the passages between anchorages are greater than can normally be covered in a day of motoring if the wind is moderate and an early start is made, and nearly everyone bound that way *does* motor. The first of these along the desert Baja coast, from San Lucas to Bahía Magdalena, is 167 miles, and for most of this we kept one foot on the beach to avoid the foul current and stronger headwind offshore. On this trip we crossed the Tropic of Cancer, so the sun chasing us north could not now overtake us. Just as though that imaginary line was something tangible, the weather at once became cooler, sweaters were needed at night, and the butter became firm. We would have liked to spend some time in the large expanse of mostly sheltered water of the Magdalena area which is alive with birds, and if one did this in a shoal-draught vessel it should be possible to make one's way through the unsurveyed channels for 60 miles to the north before having to go out into the ocean again.

The second long passage was from Cabo San Lazaro towards Bahía San Hipolito, and we and our consort left the anchorage in the lee of the cape before dawn. The sailing directions warn of a dangerous inset towards the shore north of the cape, and there was ample evidence of this when daylight revealed two merchant ships on the shore. The one nearest the cape looked to be in good condition and had presumably not been there for very long; she lay behind a spit, and was so high up the beach that most of her bottom was visible. Throughout that day *Evening Star* had been within sight, but during the night we lost touch with her, and in the morning found ourselves on an empty sea with indifferent visibility. While reading up Bahía San Hipolito we were startled to find in the latest corrections that a

◀

12. *Top*: Turtle Bay, a Mexican fish-canning settlement in the Vizcaino Desert. As there is no spring or river and never any rain, every drop of fresh water has to be brought from Cedros Island 30 miles away. *Bottom*: *Wanderer* and *Evening Star* share with the sea-lions the anchorage at the north end of Cedros. The grey line in the distance is not land but a cloud-bank over the desert coast where lies Scammon Lagoon.

shoal with only two feet over it was reported to lie in the middle of the bay. This might be impossible to see, especially if the sun was low, and we decided instead to head for Asuncion, the next bay to the nor'-west, where there appeared to be a good anchorage. We fixed our position by observations of the sun, and without difficulty reached the anchorage at nightfall. We felt a little anxious for the Maertins as we were certain they did not know about the shoal which lay directly on their course to the anchorage at Hipolito, or surely they would have mentioned it when we were all discussing going there; but our anxiety would have changed to alarm had we known what was really happening to them that evening.

Walt had called us as usual at 1630 and said that he had not yet sighted land. This message was picked up by a yacht lying at Hipolito and her skipper offered to take D/F bearings of *Evening Star* if she would remain on the air, and 'talk her in' by radio. Walt accepted, but was astonished when he was advised to alter course from north-west to north-east; however, he concluded that he must be ahead of and to the west of his dead reckoning position. There is little available information about the currents and tidal streams of this coast, but the sailing directions mention that in 1921 a U.S. navy ship, while standing by a vessel wrecked 25 miles from Hipolito, had experienced a 4 to 5 knot current which lasted for 12 hours and then stopped. No doubt Walt supposed that a wayward current such as this might be responsible, so he altered course as directed. It was after dark when he and Katie sighted land ahead, and, supposing this to be the western horn of the bay they altered course to starboard, expecting the land to end quite soon so they might round it and come to the anchorage. But the land went on and on, and presently a line of rocks and breakers showed up ahead, and close. They quickly realised that they must be off the east and not the west side of the bay; they turned 180°, and as soon as they were a safe distance from the breakers anchored for the rest of the night; fortunately the wind had died by then. It later emerged that the man who took the radio bearings had not used a direction-finding set before; he thought that the loudest part of the signal, not the null, gave the bearing, and thus was 90° in error.

Two stops later we and *Evening Star* were at anchor together in San Bartolomé, a safe harbour which is known to most people on the coast as Turtle Bay. A settlement of timber and corrugated-iron houses and shacks (Plate 12, *top*), most of them gaily painted in a variety of colours, sprawled over a sandy plain between the bare, brown hills; earth-closets stood between them like sentry-boxes, and an elaborate timber church was in course of construction near the shore, from which a long pier jutted out. About 1,500 people were living there, all engaged in the fishing and canning industry which was mostly concerned with abalone, a shell fish similar to the ormer such as one gets in the British Channel Islands. Considering that there was not a drop of fresh water to be had in this part of the Vizcaino Desert—most of Baja California is a desert with purple mountains in the distance—and that water for washing and drinking had to be brought 30 miles from Cedros Island, the people looked remarkably clean and healthy, and there was a spontaneous friendliness about them that went straight to one's heart.

An indentation, one could hardly call it a bay, near the uninhabited north end of Cedros Island was our next stop (Plate 12, *bottom*). There the almost barren 2,000–3,000-foot mountains rose steeply from the shore, and on the few, narrow, stony beaches between the cliffs were large parties of sea-lions whose barks, coughs, snorts and sighs filled the air by day and night. The bottom shoaled steeply, and although we anchored in 5 fathoms, by the time we had veered sufficient cable and had swung to it, we had 18 fathoms under our keel. The intention was that we should rest until midnight, and then start on the 80-mile passage to the next anchorage; but by midnight a strong headwind was blowing, and fishing-vessls 'up the line' were reporting bad weather by radio. So we all went back to our bunks again, and we remained at anchor there all told for 5 nights, but this delay was not entirely due to the weather.

On the second evening there the squalls off the mountains were fierce, and before turning in for the night I decided, for our peace of mind, to let out some more chain. I veered a few fathoms, but the chain then jammed in the too small navel pipe. While I was in the act of freeing it a particularly violent squall arrived; the chain jumped a couple of notches on the gipsy, and the thumb of my right hand was carried in and held between chain and gipsy, with the full strain of the cable crushing it. I had to call Susan, who was already in her bunk, to come on deck and take the weight off the cable so that I could release my thumb, which of course was badly damaged with most of its works now hanging on the outside like those of an American steam locomotive. Katie kindly provided us with up-to-date antibiotics and other items to augment those in our first-aid kit, and Walt by radio made enquiries to see if a doctor happened to be in any vessel coming our way. As a result of this publicity my thumb was often discussed among the fishing fleet, and once when we were listening-in to the morning schedule, we heard a fisherman who was talking to Walt say he would like to speak with this English guy who had bust his thumb; when did he have a 'sked'? Walt answered that Hiscock had no transmitter. There was a stunned silence for a moment, then 'Holy smoke,' came the reply, 'how in hell did he get here from England?'

The night following my misfortune the squalls were again so violent that Susan got out of her bunk to see that all was well, and to her dismay saw that *Evening Star* had dragged her anchor out into deep water and was rapidly blowing away to the south-east. She blew our foghorn repeatedly to try to wake our friends, but with no result; *Evening Star* remained without a light or any other sign of life while she grew smaller with increasing distance. Realizing that sound was not going to wake them, Susan tried light; she trained our powerful Aldis signalling lamp on their doghouse windows and started flashing. It was with immense relief that at last she saw a light appear, and very soon *Evening Star* was safely back in her original anchorage close to us.

It was unlikely that any harm could have befallen the Maertins that night, for they would have had to drift more than 40 miles to leeward before reaching the mainland, and surely would have woken up long before that could have happened. Nevertheless it was not a very inviting shore for which they had been heading. It was a desert, of course, but was scalloped by several lagoons, the largest of which is known as Scammon Lagoon, after the U.S. whaling captain who with the brig

*Boston* out of San Francisco entered it in 1857. He found that it extended in an easterly direction for more than 30 miles and was surrounded by undulating sand-dunes with no vegetation except cactus. But the important thing from Scammon's point of view was that the upper reaches were alive with grey whales, particularly cows with calves, for it is to these lagoons that this species of whale, which spends the summer in the Bering Sea and Arctic Ocean, comes to mate and breed, returning north in the spring; this is a round voyage of 8,000 miles done at an average speed of 4 knots, which is a little less than we in *Wanderer* can manage. The grey whale, which may weigh 35 tons, and is considered to be the most dangerous of all species, especially when with its young, was almost exterminated by the greed of the whalers, but today under international protection is building up its numbers. Although no whaling is now done in the lagoons, another form of commerce has sprung up: a large quantity of salt is produced, and this is ferried by barge to the south coast of Cedros Island where mechanical elevators build it into a huge heap which in clear weather can be seen from many miles to seaward; there it waits for a 100,000-ton bulk carrier to transport it to a small Japanese island, where the bargeing operation is repeated in reverse. Somehow this seemed unlikely, but one evening we saw from our anchorage at Cedros the lights of that great lumbering giant as she headed south to load her cargo.

Five days after my accident I was feeling a bit better, and the weather had improved enough to tempt us to sea. We were glad to go, not only to leave that lonely place with its constant sea-lion chorus, but because we felt we were holding up Walt and Katie, who wanted to be back in their home port by or before 29 May; but they declined to leave us on our own and were determined to escort us safely to San Diego.

We left at midnight and had an uneventful trip to the next anchorage, the sails doing a little work from time to time. But with only three good hands between the two of us we did feel rather vulnerable, and hoped that bad weather would not set in again, for we found the work on board was heavy at the best of times. We spent nights at two small islands, Geronimo (inhabited by fishermen and their families) and San Martin (with nobody living on it at that time). It was during the night that we lay there, which was not particularly windy or with visibility any poorer than usual, that the schooner *Goodwill*, which we last saw at San Lucas, struck Sacramento Reef, inside which we had passed the previous morning, and went down with the loss of all of her 12 hands. On arrival at San Diego we were cross-examined by the coastguard and a private-eye employed by the owner's family, as it seemed that we and *Evening Star* were the only other vessels in the neighbourhood at the time of the schooner's loss. Mystery surrounded this tragic happening. Why did the lookout not see the breakers on the reef or hear their roar, for they were to windward? Why did he not see the 19-mile light on Geronimo, which was only 5 miles away, and was flashing brightly enough the night we were there? Why, even if the schooner sank immediately, which seems unlikely with so big a ship, did nobody cut the lashings of the boats? All three were reported as having been seen from the air to be still on the deck of the sunken vessel. Two bodies were recovered soon after, one wearing pyjamas over swimming trunks, and a month later two more were found in a lagoon 25 miles south-east of Sacramento Reef.

After a hard punch against a rising wind and an awkward swell we anchored for one night only, we hoped, in Bahía Colnett, which in the conditions prevailing was just about as desolate a spot as could be imagined. We brought up off a remarkable gorge in the lee of the flat-topped, semicircular headland with vertical cliffs, which sheltered us from the steep sea but not from the swell which, measured by our echo-sounder, had a height of 7 feet at the anchorage. It seemed that we were on rock, so we shifted berth, but with no improvement, and the rumbling and rasping of the chain each time a squall hit us was horrid to listen to. Fearing that the cable might catch on the rock, snub short and possibly part, Susan kept a half-hourly lookout through the night, for astern lay the curve of the bay with its shoals and breakers. It was some comfort to her to see *Evening Star*'s riding light bright and close. I should have shared that miserable task with her, but my mangled thumb was making me feel far from well, and at times even far from caring. We left before dawn, but having rounded that seemingly interminable headland, we found conditions too rough to proceed, so both yachts returned to spend another day and night at the uneasy anchorage, where, to add to the general feeling of apprehension, a whale of considerable size (could this have been a straggler from Scammon Lagoon?) swam close around, sending a sighing spout aloft now and then. Even for south-bound vessels the weather was too bad, and two came in to shelter with us. Later Walt and Katie gave us one of those little plaques of which American yachting people are so fond; it was to remind us of our time in Colnett, and the inscription read: 'Sailors have more fun'.

It was indeed a relief to get away next day, and our joint plan was to spend part of the night at the Islas de Todos Santos (All Saints Islands) which lay on our course for San Diego and about 10 miles from the port of Ensenada, and then to hurry on to San Diego, hoping to arrive there before the Memorial Day long week-end holiday commenced in two days' time, or we would have to pay a high overtime fee to the customs. We reached the islands after a long day of motoring, but found the anchorage so uneasy with the swell rolling in that we went across to the well-sheltered harbour at Ensenada for a few hours of sleep, and then continued on our way.

That day for once was calm, and we made excellent progress along the coast, cleaning and polishing away the stains (mostly rusty) of our long battle against the wind and sea, and we ran up clean new flags. As we crossed the border between Mexico and the United States we seemed suddenly to come into another world. Overhead jet aircraft and rattling helicopters shattered the silence which had been ours for so long; a freighter with a bone in her teeth kept company with us for a bit, then drew ahead, and a pair of submarines overhauled us; sport fishermen with high-flung flying bridges and creaming wakes cut so close that we feared collision, and small holiday boats, with mum, dad, and the kids aboard criss-crossed our course.

Walt had been busy with his radio, and when a little in advance of our E.T.A. we crossed the main traffic lane, which was rather like crossing a busy street except that the traffic did not move quite so fast, and went alongside the boarding dock on Shelter Island in San Diego harbour, the officials came almost immediately to enter us and seal the bonded stores locker, in which were the remnants of the liquor we had shipped at Grenada (later they let us break the seal and help ourselves). An attractive and efficient young woman then stepped aboard and piloted us across the harbour, in which upwards of 2,000 immaculate yachts lay at their floating finger piers, to a guest berth right in front of the imposing San Diego Yacht Club. People were waiting there to take our lines, and the moment these were made fast, thick-set John Bate with the horn-rim spectacles and the iron-grey hair, put me into his Volkswagen van almost by force, drove me through the traffic-loud express-ways and streets where the neon signs were beginning to flicker, and within half an hour of docking I was in the hands of Dr. Stephen Thein, a yachtsman. Later that evening, after he had performed an operation on my thumb, which X-rays showed to have the bone broken into four pieces, he drove me back to *Wanderer* and Susan. Having started at Ensenada at 0330 this really had been quite a day, and I settled into my bunk with the lovely feeling that we were no longer to worry about anything —we were in capable, almost overwhelmingly kind and hospit-able hands.

# 4

# A Year in California

The fairway of San Diego Harbour has to be dredged from time to time to keep the depth to 40 feet for the large naval and merchant ships that use it, and for many years the dredgings were taken out to sea and dumped. But then John Bate became port director, and he had a better idea: he reckoned he could reduce costs and at the same time create some real estate. So he arranged for the dredgings to be dumped on a shoal in the harbour just in front of the suburb of Point Loma, and close to the beach where the brig *Pilgrim* (in which Dana served) and other ships engaged in the trade loaded hides in the 1830s. In time the shoal rose above water and was given the name Shelter Island because it formed the south-east sides of a yacht harbour and a commercial basin, and protected them from any swell that might run up the harbour, and from the wash of passing ships. When the one-mile-long island had been formed and joined up with the shore, restaurants, clubs, marinas and motels were built along its northern edge; but the rest of it was sown with grass and planted with trees, and it with its beaches was declared a public recreation area. Bate's Folly, as Shelter Island in its early stages was called, proved to be a money spinner for San Diego, and after its completion John was invited to make some more real estate out of the mudflats, which he did.

Over the years Susan and I have made use of many yacht marinas, the word implying a collection of sheltered and secure berths alongside (usually) pontoons so that one may step ashore at any state of tide. A few were so squalid that we imagined our ship might catch some disease, or at least have rats on board; some were so expensive that we worried about the pounds or dollars clocking up; and occasionally one was so perfect that we could relax in it with pleasure and enjoyably make use of the facilities provided—the marina of the San Diego Yacht Club was one of these. The handsome club-house, with its high-peaked roof, stood with its front legs in the water of the Point Loma Yacht Harbour; the gangways leading to the complex of walk-ways from which the pontoons (or docks) sprouted, were thickly carpeted so that no dirt should be carried aboard the yachts; there was water and electricity at every berth, there were little rubber-tyred carts for bringing stores from the car-park, and by night the place was gently lighted. Each walkway had its telephone, so when one of the girls with the golden voice in the office said over the public address system: 'Mrs. Hiscock, telephone please'—and this happened frequently, for rarely did a day pass without some kind person inviting us out—Susan had only a few yards to go. The club provided hot showers and a restaurant with excellent food, and within a hundred yards or so the shops, banks and post-office provided all our other needs. At least once a week Katie drove Susan to the laundromat, and while the washing rotated in the dryers they got all the stores they needed from the adjacent supermarket.

Life was almost too easy and comfortable, and we felt that if we remained too long we might grow so fat, soft and idle that we would never want to go to sea again. Even the weather played its part. There was no rain at all for the first six months and very little after that, so that outside painting or other work could be done on any day; sometimes the mornings were grey with mist or fog, and this happened more often as the summer advanced, but by 0900 it had usually scaled up and the sun shone for the rest of the day; a nice breeze made at noon or soon after, so those who wished to do so could have a little sail or a race; by nightfall the wind died, and the lights of the clubs and houses surrounding us gazed at their unwinking reflections in the mirror-smooth water, and at Christmas time, when many of the buildings and some of the yachts were festooned with strings of coloured lights (Plate 13 *top*), the effect was exciting and attractive.

With easy minds we could leave *Wanderer* at any time, for there was a watchman at the gate (often by night he could be seen picking over the contents of the marina garbage cans in search of something useful), and a police launch patrolled the harbour to keep an eye on everything, and to ensure that the 5-knot speed limit was not exceeded. Although *Evening Star* lay in the same marina as we did she was some distance away, so when we dined aboard her we usually rowed round in our dinghy as Nicholson preferred this to the long, scarey walk. One night we were rowing home, seated on the middle thwart with an oar each, as is our custom, with Nicholson in the bows keeping a lookout, when the police launch overhauled us.

'Say, d'you have a flashlight in that boat?'

I mumbled something about coming from England where we do not usually carry lights in dinghies, and felt very small when the courteous reply came back:

'Well, a flashlight might be a wise thing to have in any country. Good night.'

The generosity of the big-hearted club was remarkable, and throughout our long stay it continued to treat us as honorary members, and all its facilities were at our disposal. How fortunate we had been to fall in with the Maertins, who were old members, and to have been introduced by them! They also introduced us to their own special little circle of friends, generally referred to as the Gang (Plate 13, *bottom*). Among these were Ray Quint, an ex-marine who lived a bachelor existence in his ketch near us, and who did some good organizing so that we were able to visit in the most enjoyable manner such California tourist attractions as Disneyland, the best amusement park in the world (Plate 17, *top*), and the 200-inch telescope in the Mount Palomar observatory. Across the harbour at the Silver Gate marina Barbara Cochran and Maxine Bleming lived aboard their *Island Queen* in which they had made some fine cruises up and down the coast. They called themselves Captains Grimjaws and Cementhead, and I felt embarrassed when Barb, who had piloted us across from the boarding dock to the yacht club, asked me to guess which of these unbecoming names belonged to her. With the Gang we had many a 'happy hour' when the bourbon flowed and Walt fondly embraced whichever female happened to be within reach. There were many other delightful people, such as Barney and Dolly Bell in the beautifully fitted motor yacht *Dos Campanas* (two bells) which among other rich appointments had gold-plated fittings in the bathroom. Barney, tall, slim and courteous, had in his younger days been a stunt man in the Keystone Cops films, and it was in Hollywood that he met his charming Dolly, for she had been in movies, too. They were our kind and understanding hosts when, with our bare toes buried in the deep-pile carpet of their large deck saloon, we watched on their colour television the first landing on the moon, and saw the film featuring the Royal Family. Most of the other people we got to know also showered hospitality on us, just as though we were movie, t.v., or sporting stars, whereas so far as we knew our only singularity was that at the time ours was the only yacht flying the U.K. ensign on the coast of California; and many an evening we spent as their guests in one or other of the clubs or restaurants which ring the harbour. Whether it was just a passing fashion or a facet of the American way of life, we did not discover, but in public eating places one had to dine almost in darkness, as the only light was shed by one or two flickering candles on each table—for reading the menu that flashlight the police mentioned might certainly have been a wise thing to have. It was said that twisted ankles were an occupational hazard of dining in these establishments. Very much more enjoyable for us were the meals we ate in the well-lit and well-appointed homes of our friends, as, for example, those of the Bates (John and Sylvia) or the Theins (Steve, my kind doctor, and his wife Bev). There were some other homes whose owners were not quite so well known to us, and it was always interesting to see where some of the money went: one had parquet flooring, but this was not of wood, it was of pigskin; at another the windows were thermostatically controlled so that they all opened and closed automatically as the temperature changed; a third had a room lined with bins in which a fabulous collection of bottles of vintage wine was stowed, but it was clear that the labels on the bottles, not the contents, were the interest, for definitely the wine was not for drinking; one house was so large it covered half an acre.

On my early morning, pre-breakfast walks I often passed a house in which we had been entertained; grass was mown, miniature hedges trimmed, flowers blooming, and sprinklers, fed no doubt by water from the long-suffering Colorado River, were sending their life-giving spray over all. The centre of the town was not so glamorous; in the empty car-parks oily patches lay between the neat white lines like droppings in a milking bail, and with whirring machinery the great trucks of the garbage-collectors were digesting the previous day's waste. I enjoyed reaching the summit of my favourite, breath-taking hill, which was so steep that its road had been sealed with cement instead of asphalt, and pausing for a moment to look down on the quiet water of the harbour with its yacht-laden marinas, and, if there was no fog that morning, out across Shelter Island to the slender and rather lovely skyscrapers of downtown San Diego (Plate 14). By the time I had descended my hill the early morning rush had begun with a stream of cars with still cold carburettors and smoking exhausts banging and roaring along Rosecrans towards the naval base, and there were joggers with whom to exchange 'Hi'.

As I cooked breakfast, which I always enjoyed doing while Susan attended to her long, auburn hair, made the bunks and tidied up Nicholson's heads, I used to listen to the news on the radio, and usually was shocked to learn that some act of violence—murder, arson, rape, or race-riot—had occurred in the near-by city while we in the quiet, policed citadel of the well-to-do at Point Loma had been enjoying our sheltered, pampered existence. But it was difficult to concentrate on the news because of the commercials which punctuated it:

'With Colonial Ford you always know right where you are'—'Alan John for the executive look . . . the well-known brand of the well-dressed man'—'San Diego Trust has good things going for you: the check-guard cushion to stop checks bouncing'—'Rancho California's sophisticated wilderness'.

However, I did glean items of news from between the advertising jingles:

The massacre at My Lai where, it was said, 109 Vietnam civilians were killed by U.S. troops, and 26 soldiers stood accused. First landing on the moon . . . 'One small step . . .'—Operation Intercept, an attempt to stop the smuggling of marijuana across the Mexican/U.S. border. This seemed to me a hopeless task, for the border is 2,500 miles long, and a 1·8 kilo brick (gift-wrapped at Christmas) which in Mexico costs $35 and is enough for 2,000 cigarettes, or joints, was selling in the States for up to $1,900. The tiny force of 40 Mexican drug agents were so poorly paid that they were easy prey to *mordida* (the bite, or pay-off). Presumably the operation could have had no more success than prohibition had, for there are places on the border where a small army could cross undetected. Very soon came Mexico's retaliation—a medical check was to be imposed on every U.S. citizen crossing the border. Fortunately for us all this happened just after friends had driven us for a day of exploration in Mexico. Meanwhile the eggs got boiled, scrambled or fried, or the thin-sliced bacon turned into golden, curly crisps in my pan.

Living aboard in a marina does have certain hazards which chiefly revolve round one's neighbours, and the nearest of these may be no more than three feet away across the dock. Too loud and frequently used radios come high on this list, along with people who charge their batteries noisily and fill one's ship with exhaust fumes. The man who whistles while he works or mooches around, or entertains his guests on deck with too much hard liquor. Boys who fish and leave their bait and hooks to foul one's feet, dogs which use the miniature lamp-posts, and little girls who run to and fro on flat feet shaking the whole establishment. At San Diego Yacht Club we suffered from none of these things, for not many people lived aboard there, and those who did were considerate of their neighbours; neither were we bothered by wire halyards tapping alloy masts at night, for there was never any wind then. However, we did have one small experience which is probably unique.

One morning I was working in the blacked-out forepeak making enlargements with which to illustrate a story I had written, while Susan had gone ashore to shop with Katie. I was getting along well when something metallic started at irregular intervals to strike *Wanderer*'s deck and topsides near the bow. For a little while I tried to ignore this interruption, but eventually curiosity and the thought of possible damage to the paintwork got the better of me, so I slipped the last print into the fixing solution, closed the box of bromide paper, and went on deck to investigate. Coming out of the forepeak, where the only light came from a dim orange globe, I was dazzled by the blazing sun, but so far as I could make out there was an enormous black bird hovering overhead; now and then it lowered a wing-tip and gave us a swipe. As my day vision improved I saw that this was a black nylon spinnaker flying horizontally from the masthead of the yacht lying next to us in the breeze which had sprung up a little earlier than usual. Apparently the owner had been airing it when the tack carried away, and now it was the thimbles at the tack and clue that were belabouring us. I got our long boathook, and when in a lull the sail descended to within reach, I hooked a corner and thrust it towards the owner: he thanked me and I at once returned to my work. Susan got back to the marina about then, and watched with dismay as the sail lifted once more and then flopped down on to one of the piles which was shod with an anti-seagull spike, where it got badly torn.

A day or two later I happened to meet the owner in the club and I commiserated with him over the damage to his beautiful black sail. This was the day before the new 2-mile-long bridge spanning the harbour between San Diego and Coronado was to be opened. The occasion had received much advance publicity, and all sorts of people were to take part in the first crossing, including cyclists and joggers. Jokingly I asked the owner of the spinnaker if he would be one of those jogging over the bridge in the morning. He gave me a long, straight look and said:

'No, I shall be dedicating the bridge tomorrow.'

Perhaps I misunderstood him, but I failed to discover whether he was the bishop or the governor.

Our only real problem while lying at the marina was that we had a great many visitors. We like meeting people, and this is one reason why we go voyaging, but we were not getting enough privacy in which to attend to our own affairs. We found that ten to twenty visitors a day were too many, especially when most wanted to come aboard, talk boats, ask questions, and stay sometimes for several hours. Frankly, we were astonished that there were so many people with, apparently, nothing else to do. Most of our visitors were dreamers, and some would never get themselves afloat because they had (or thought they had) too many important ties. But they sure did want to talk, and some even started to argue when Susan or I, having been asked a direct question, such as what did we think of roller furling head-sails or why didn't we cook by gas, had expressed an opinion: 'But Irving Johnson has a roller jib'—'Gas is not so dangerous as you folks seem to think.' John Bate, realizing what our trouble was, kindly put his mooring at our disposal. This lay only 30 yards or so off the end of one of the club walkways, and lying on it had the merit that anyone who really wanted to get at us had either to borrow a dinghy (there were not many of them around) or else hail us to fetch him—incidentally, the American way of hailing seems much better than the method we use; they shout 'Aboard the *Wanderer*', the 'aboard' alerting one ready to listen for the name; we hail '*Wanderer* ahoy', and the name, coming first, may often not be properly heard. Life then became easier for us; also we felt we were not outstaying our welcome at a club guest berth.

After two and a half months at Point Loma we thought we ought to move on before the urge to do so was killed by the soft life we were living; also it had always been our intention to spend the winter months in the San Francisco area. I particularly longed for the invigoration of a spell at sea as an infected bone in my damaged thumb had kept me on antibiotics for a long time; but now Steve, who had remodelled the thumb so that later I was to have almost the full use of it again, pronounced it sufficiently seaworthy provided I was careful with it. When we told our friends that we were leaving their reactions were much as we had expected them to be.

'Reediculous,' said Ray, 'you must be some kind of a nut.'

'Are you kiddin'? asked Katie. 'San Francisco's no place to spend the winter; too cold and wet. You'd do far better to stay right where you are.'

'It's the wrong time of year for heading north,' Walt reminded us, 'with lots of fog from Conception on. Of course a few yachts do it, but they use their fathometers all the time to keep inside

▶

13. *Top*: At Christmas time the San Diego Yacht Club and some of the yachts at the marina were festooned with strings of coloured lights. *Bottom*: Happy hour with the 'Gang'. Katie Maertins is next to Susan, Walt has one arm (his right) round Maxine Bleming and the other round Barb Cochran. Ray Quint is at the right.

the shipping lane. Yours doesn't seem too good in deep water, and before you know it you'll be out among those tows. I expect you heard of the power yacht man from Newport who in fog passed between a tow-boat and the barge she was towing, and saw neither—but the towline caught him and took his flying bridge right off. No, you'd best stay right here like Katie says.' His hatchet face crinkled with an infectious smile. 'Besides, we like having you around.'

However, we felt we must go, so after a round of farewell parties we left the harbour, where we had been so well looked after, one quiet evening when the visibility was rather better than it had been for several weeks.

Some of the greatest concourses of yachts in the world are to be found in the (mostly artificial) harbours which lie along the 180 miles of Southern California coast between Santa Barbara and San Diego; in Marina del Rey harbour alone there were said to be 8,000. There is nowhere much to which all these yachts can cruise at the week-ends except the Channel Islands, and our intention was to have a quick look at one or two of these before heading north for San Francisco. There are eight islands in the group, and it was for one of the nearer ones, Santa Catalina 60-odd miles away, that we steered as we left astern our good friends and the bright lights of San Diego.

The afternoon wind had died at sundown, as it usually did, and we motored through the night barely making 4 knots when we should have been making 5, and were puzzled by this. Shortly before leaving Susan had dived down to give the bottom a scrub, and so that she could hold herself underwater we had rigged a line from rail to rail under the keel; when at dawn I went aft to read the log and find out how far we had come, I discovered that we had forgotten to take the line in, and in the meantime it had collected a big bunch of kelp, of which there is much along this coast; we went a bit faster after we had removed it. A little later an aircraft carrier with many planes on her deck

◀

14. San Diego from my favourite early-morning hill. In the middle distance stands the high-peaked roof of the yacht club with its marina spreading round it. Beyond is the palm-fringed strip of Bate's Folly, properly known as Shelter Island, a piece of real estate made from harbour dredgings. Five miles away stand the sun-gilded skyscrapers of down-town San Diego.

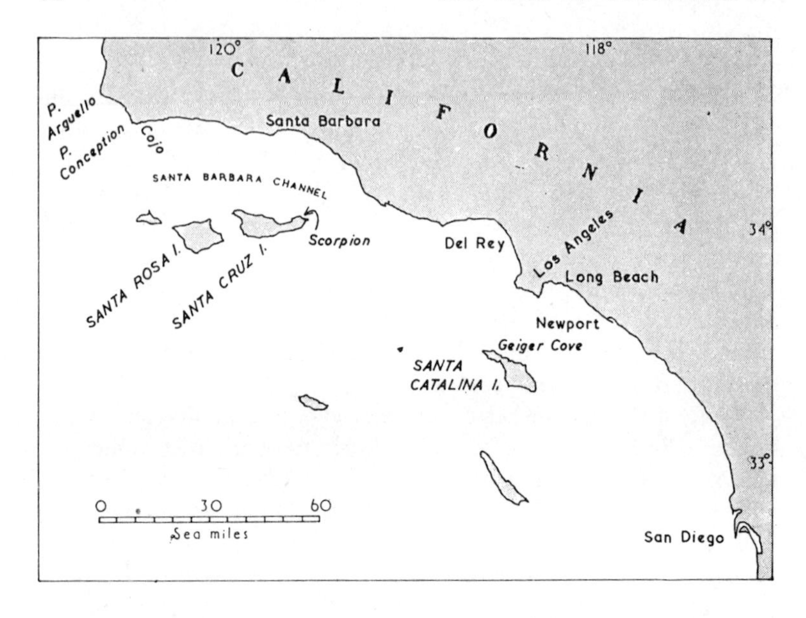


steamed majestically by; an umbilical cord connected her to a depot or supply ship, and astern, as though ready to pick up anything dropped, slunk a long, lean, handsome destroyer.

Santa Catalina, which is about 18 miles long, provides the most frequented stops for week-end yachts, and most of its more sheltered bays are thickly sown with moorings. Some of the bays are leased by mainland clubs, and only members are welcome in them. However, a friend who belonged to the Blue Water Club, which leases Big Geiger Cove, had invited us to call there as his guests, so we headed along the island's north coast towards it, passing on the way the holiday town of Avalon, where a mass of yachts lay moored with ferries and speedboats weaving creaming wakes among them, and aircraft buzzing overhead. Our friend's yacht was one of the 'character' type, with a clipper bow, a springy sheer, and upflung bowsprit which almost blocked the helmsman's view, and she was painted red and white; so we could pick her out at a distance from among many other craft, and thus had no difficulty in identifying Big Geiger Cove, which is not too easy to spot from the offing. She and her close-packed neighbours had stern anchors laid out towards the beach, but as we intended to stop only for a few hours, we lay to a single anchor on a kelp-foul bottom just outside them, and at once had visitors.

While motoring along the coast that quiet forenoon one of the bolts holding the electric stopping device to the engine's cylinder block had broken, and I was hard put to it to keep the engine running properly. It seemed unlikely that in this secluded spot we would be able to obtain a replacement bolt of the right size, or help in fitting it if we did; but when we mentioned our trouble one yacht immediately produced from her crew an engineer/electrician, another provided the bolt we needed, and a third lent us some sophisticated tools. Within an hour the stub of the broken bolt had been extracted and the new bolt fitted, and we were then socially engaged for the rest of the day.

0200 a nice little west-sou'-west breeze tempted us to sea, and having weighed with some difficulty because of the weight of kelp clinging to chain and anchor, we set a course for the channel which separates the islands of Santa Cruz and Santa Rosa, some 80 miles away. The night was overcast and very dark, the wind soon fell light, and before dawn fog set in so that we could see nothing but the reflected beams from our sidelights. However, after daybreak we sighted in a temporary partial lifting of the fog the misty cluster of little islets off the southern point of Santa Cruz, and then groped our way into the channel. At noon the wind veered and freshened and blew the fog away, and we found a good anchorage in the lee of Santa Rosa off a pier which serves the ranch there. This channel had a bad reputation, for the north-west wind which generally prevails blows strongly through it.

By now I was beginning to find that my damaged thumb was not yet fit to do much work, and I was incapable of hauling strongly on a rope or even of getting about the deck safely. So as next day the wind remained ahead and fresh, we stayed at our pleasant anchorage, and with amusement listened-in morning and evening to the weather summaries from the San Pedro marine radio operator. I think she must have been a rather simple girl with little understanding of what she was reading. For 30 knots she said 'Three zero kilo November.' For one-half she said 'One slant-bar two.' Shortly afterwards she was taken off that job and the summaries and forecasts were read by a coastguard with great improvement, but we still found it difficult to understand why one place might have 'Visibility 4 miles in fog,' and another 'Visibility 1 mile in haze.'

Every coast has by local repute its own Cape Horn: Ardnamurchan in Scotland, Hatteras on the east coast of the U.S., New Zealand's Puysegur Point; and although these are weaklings compared with the real stormy capes, such as Horn, Agulhas, Leeuwin, Farewell, they do provide something of a challenge. California has its Point Conception, and our hope was to round it next day.

We were up and breakfasted before it was light, but the fog was thick and we did not get away until 0600. To start with there was just enough wind to keep the sails asleep as we headed across the Santa Barbara Channel, but later it became strong, and again I found my injured thumb a handicap; so we made a slight alteration of course for the Cojo anchorage, which lies a little to the east of Point Conception. This is a strange little spot in an indentation of the coast, and is protected to some extent on its seaward side by a bed of kelp. There is no chart of it, but Barbara Cochran and other friends had told us about it and made little sketches for our benefit, and having entered through a channel in the kelp, we turned to port and brought up as directed off the culvert which pierces the giant railway embankment that skirts the shore.

The wind took off in the evening, and no doubt we ought to have put to sea, but instead we turned in and had a peaceful enough night except for the rumble of trains—some of these comprised 100 trucks drawn by three diesel locomotives. We weighed at first light, when we could just see and avoid the kelp, but immediately fog rolled in and we re-anchored. The 0800 weather report from Point Arguello, which lies a few miles north-west of Point Conception, included the words 'Visibility one slant-bar one six,' which we correctly interpreted as 1/16th of a mile.

Throughout the forenoon we sat and shivered in the chill damp (how foolish I was not to have rigged the chimney and fired up the cabin heater), while we listened to the trains rumbling by and read Dana's *Two Years Before the Mast*, which was not very encouraging, and did the homework which we ought to have done before starting on this trip. There are several yachtsmen's guides to this coast, some of them big, superbly illustrated, and very expensive, but much the best that we came across was the Corps of Engineers' *Small Boat Harbours and Shelters* (unfortunately it is now out of print). This gives a description and a colour chartlet of every possible anchorage between San Diego and San Francisco, and straightforward information about the weather likely to be met with. It told us that July to September are the worst months for fog (by now it was the third week of August) and that the fog gets worse as one goes north from Point Conception, often lifting only from noon to 1700. Sure enough it did lift at noon, but then the headwind, even in our sheltered berth, piped up to 25 knots and I did not feel equal to coping with it.

The following night I looked out at 0200. There was no breath of wind, but it was thick as a hedge again, and cold. I climbed back into my bunk, and Susan and I had a talk, the upshot of which was that, apart from the obvious difficulty of getting there with only three good hands between the pair of us, San Francisco would be a damp and chilly place in which to spend the winter (Walt, Katie and Ray were absolutely right), and that we would do better to return to warmer San Diego, where in almost guaranteed good weather we could attend to the many jobs that still needed doing on board in preparation for our forthcoming Pacific crossing. I believe this was the first time we had ever abandoned a major project, and it didn't make us feel too good; but our decision proved to be a wise one; for the next eight days the reported visibility each morning at Arguello never exceeded half a mile, and often was zero, while headwinds of 35 knots were reported in the afternoons.

At dawn we motored away in a calm with very poor visibility and passed close to several of the brightly lighted oil-rigs which stand on stilts in a line a mile or so off the coast, but were careful to avoid the area off Santa Barbara where for months a leak had been spewing out tons of oil to foul the sea and the beaches and kill the wildlife. We steered for the Island of Santa Cruz, and as visibility improved to a little over a mile as we drew away from the mainland, we had no difficulty in finding it. On its north shore we looked in at Pelican Cove, a popular anchorage, but found it so crowded that we bore away and ran on down to the Scorpion anchorage and brought up in the lee of some islets where seabirds nest. Perhaps because we were a little early in the day no other yachts were there, but soon we had seven neighbours; the sun shone warmly and we watched the people, mostly families, from the small yachts surrounding us having a wonderful time swimming, fishing, and exploring the channels and the dark caves that penetrate the cliffs; Cojo with its chill fog and tearing wind, though only 50 miles away, might have been in another world.

Marina del Rey on the mainland coast near Los Angeles was our next stop; it is said to be the largest marina in the world, and fills its own harbour, which was dug out of swampland. Inside we were reluctant to go to the California Yacht Club because it lay down to leeward, and if it could not accommodate us we might have difficulty in getting away again. While we were jilling about trying to decide what to do, a young bikini-clad couple in an outboard dinghy circled us and shouted: 'We have just been reading your Atlantic book.'

'Well,' I remarked to Susan, 'what a moment to choose to tell us that.' But she, more alert than I, asked if they thought we might go to an empty berth at the Del Rey Yacht Club which lay to windward—most clubs along that coast have guest berths for the use of visitors from other clubs.

'We'll soon find out,' and away they sped, bikinis and nothing else, into that rather austere establishment, and shortly returned with an invitation. So we berthed there where the pontoons had rubber wheels at their corners to prevent damage while berthing, and the piles were surmounted by pots of flowers, and we had a magnificent motor yacht each side of us. The club was kind, and so were the people from one of our neighbours; they took us for the first time in our lives to an ice-cream parlour where we sampled various kinds including the new Moon Mix, and out to dine. Our other neighbour had cost, we were told, $350,000, and we could well believe it. A notice at her gangway read: 'Deck shoes or stockings only,' but Nicholson was not wearing either when he decided to go aboard, and neither was Susan when she went in hot pursuit to find him sauntering with tail erect on the lime-green carpet of the deck saloon. We found this vast marina, with its thousands of plastic yachts extending as far as the eye could see, rather frightening. Each lay at her own dock for which her owner was paying about $2 per foot of overall length a month. Very few were inhabited and we were told that only a small proportion of them ever left their berths. The reason for this is partly because for some people a yacht, particularly a fully-powered one with all possible electronic equipment, is a status symbol, and partly because there are so few good places to go to that most owners prefer to do their week-end yachting at the dockside with the convenience of laid-on telephone, electricity and water, and a restaurant and car-park near by. We discovered that most yacht harbours along the coast were already filled to capacity, and although some were being enlarged and new ones built, they could not keep up with the streams of boats pouring off the factory production lines. Where, we wondered, could the new boats find berths, where could they go, what would be the end of it all?

We made the mistake of entering our next big harbour, Long Beach, on a Sunday morning during rush hour, when traffic was pouring out between the long, parallel breakwaters. Mostly the yachts were under power and were competently handled, but there were, of course, flutters of sailing dinghies stoutly maintaining their right of way, and zooming in and out among this throng were crazy people water-skiing. The Long Beach Yacht Club has no guest berths, but we were allotted a dock in one of the near-by marinas for which the club hospitably insisted on paying. Here again plastic platoons surrounded us, but we put the outboard motor on the dinghy and left the sterile scene for a little while to visit Naples, a high-class residential district where individualistic and often quite lovely homes lined the network of curving, man-made canals, each with a yacht berthed at the bottom of the garden.

After calling at Newport, which we thought the most attractive of these harbours, and where many a Hollywood star had a waterside home, we returned one quiet evening to San Diego at the end of our 400 mile potter round the bay, and as we approached the yacht club John and Sylvia Bate hailed us:

'The mooring's all yours, do use it.'

So we secured to it, and very soon were enjoying a 'happy hour' with our friends. Although they were polite enough not to say so, it was clear that some of them thought my thumb was just a convenient excuse, and that in fact it had been Point Conception which defeated us. For that reason alone we would certainly have to try again to reach San Francisco, and anyway we wanted to have the satisfaction of sailing our own ship in through the Golden Gate, but that could now wait until the spring.

Sometimes we thought that San Diego was becoming a port of lost causes. For many people it is the last stop before setting out on a trans-Pacific voyage, and most of these had come down the coast, some from as far as Canada. But on the trip south they had been within sight or easy reach of the coast, where there were many good stopping places, the lights and the radio beacons were reliable, and for anyone with ship/shore radio the coastguard could always be called on for help in the event of trouble, and weather forecasts could be obtained several times each day. So without much trouble, apart from fog, they reached San Diego to ship stores and make final preparations for the long voyage out to Galapagos, Marquesas, Tahiti or beyond—a dream to be realized at last, probably after years of saving and planning. Then the great day arrived, and they put to sea. But a couple of hundred miles offshore they found themselves out of broadcasting range, so they could no longer call the coastguard if anything went wrong, and weather forecasts for their area were no longer available; for the first time in their lives they found themselves alone on a wide and empty ocean, with the need to depend entirely on their own skill, knowledge and resources for everything, and this was rather frightening. We farewelled some of these people with good wishes and a bottle of wine to be drunk after the first 1,000 miles had been made good, and were sad to see them return to the anchorage, their dreams of far horizons dispelled, their ships for sale.

I remember a pretty little topsail schooner down from British Columbia, manned by an efficient and likeable young couple, who had made enough money chartering to take off on the grand tour—San Diego, Marquesas, Tahiti, Hawaii, and return to B.C. in time for the next chartering season. As they delayed in San Diego with visits from friends and parents, the weeks ran on into months; after a while they said they would not now be able to stop at the Marquesas and would have to go to Tahiti direct; but more time went by until it was too late for anything except the trip to Hawaii and back to B.C. Finally they departed, but four days later they were back and their ship for sale. 'There was no wind out there, and ours is not a motor vessel.' Another yacht returned after a few days of moderate headwinds. I at once rowed over as I feared someone might be ill and need a doctor to be brought out, only to learn that the small daughter had been seasick—'After the first two days it was not fun.' That yacht, too, was put up for sale.

If only these people could on the way down the coast have headed offshore for a day or two to find out what it is like to be alone at sea, I believe there might not have been so many disappointments, and certainly if yachts did not carry radio transmitters there would be fewer abortive or unnecessary air/sea searches made; and these, because of the publicity they get in the press, where it is usually stated that they have cost the taxpayer a large sum of money (probably they have cost little or nothing, as the planes would fly and the ships manoeuvre on exercises whether or not there was a yacht to look for) give the cruising fraternity a bad public image which is not warranted; for every yacht that gets herself in the news there are scores, probably hundreds, going efficiently about their business with no fuss or publicity. In the list of lost causes multihulls ranked high, so it was good to see the Canadian trimaran *Tryste II*, built and skippered by Ernest Haigh and crewed by his wife and four daughters (Plate 15*C*), set out with quiet confidence. She made the Pacific crossing efficiently and apparently enjoyably.

Our remaining time at San Diego passed quickly, for I was working on the revision of a book and Susan was attending to some of *Wanderer*'s many needs; occasionally we joined other members of the Seven Seas Crusing Association over at Silver Gate to help put the *Bulletin* together under the able direction of Larry and Babe Baldwin (Plate 15*D*), who at that time were the unpaid volunteer editors. We travelled about by air to give talks and show our film *Beyond the West Horizon*, Christmas came and went, and all too soon we were making our final preparations for departure. We hauled out at Kettenberg's efficient yard, and after scraping an astonishing growth, some of which looked like coral, some like thick spaghetti, off the bottom, we antifouled it. The last of our jobs was to check the compass, which we were able to do conveniently on John's mooring, taking bearings of the disused but conspicuous old lighthouse on the skyline, and it was purely by chance we discovered that half a turn of the wheel moved the compass card several degrees; the Dutch adjuster had indeed been right when he said there was something magnetic in the bevel box of the steering gear immediately under the compass. I was reluctant to take the box to bits for fear of damaging the oil-seals, which might not be readily replacable, and the only alternative was to raise the compass 5 inches so as to lift it out of the magnetic field of the box. This we did, and found the increased height made it easier to use the compass for taking bearings (in a steel vessel one cannot use a hand-held compass for that purpose); but it had the grave disadvantage, as we were to discover later, that because the compass is of the grid type with which, to steer a course, one keeps a line on the card parallel to other lines on top of the bowl, a parallax error could occur, particularly on courses with a large east or west component, when in high latitudes the card was pulled out of the horizontal plane by the force called dip. With the helmsman's eye well above the card this error does not occur, but now that we had raised the compass his eye was at a smaller angle to it, and the error could be considerable.

Once again, and this time with great sadness for we knew it would be for good, we took leave of our San Diego friends, and towards the end of March, 1970, put to sea. This, as Walt and Katie had told us, was much too early to think of going north, but our reason for choosing to leave then was that there should be less chance of encountering fog which, as the following quotations from the *Pacific Coast Pilot* suggest, is a considerable hazard that increases in frequency with the approach of summer.

'The entrance to San Francisco Bay is a region of frequent fog, and shipwrecks have been numerous there. Often a sheet of fog forms early in the forenoon off the bold headlands of the Golden Gate, and becomes more formidable in size as the day wears on.' Again: 'On summer afternoons the velocity of the wind at San Francisco, with almost clockwork regularity rises to over 19 knots, and a solid wall of fog comes in through the Golden Gate, causing a fall in temperature.' Of Point Reyes, which is not far from the Golden Gate, the *Pilot* says: 'This is often spoken of as being the actual center of heaviest and most frequent fog on the Pacific coast. Owing to the persistency of the fog cover, through which the sun's rays sometimes fail to penetrate for three or even four weeks at a time, Point Reyes has close to the lowest midsummer temperature of any observing station in the U.S.'

As before, we made an overnight trip to Santa Catalina Island, but as this was not the yachting season we had an anchorage almost to ourselves. Once again we put in for a night at the strange little Cojo anchorage inside the kelp beside the railway embankment, where on our previous visit we had shivered in the fog, but this time the weather was fine. The following morning, while we were trying to beat round Point Conception against the usual fresh headwind, Susan who was at the helm suddenly said: 'The steering seems to have jammed, I can hardly move the wheel.'

It took me several minutes to discover that a grease-gun, which I had failed to stow properly in the engine-room, had moved and inserted its tough rubber penis between the chain from the auto-pilot's motor and the sprocket which it drives on the steering gear. Fortunately the chain had a split-link, and after a struggle I managed to release this and extract the grease-gun so that hand steering again was possible; but I feared that the strain might have bent the motor's shaft, and certainly re-alignment of the motor and adjustment of its chain would be necessary. I felt that task called for smoother water than we were likely to find in the open anchorages near by, so we bore away and ran back 30 miles to a swell-free anchorage in the lee of Santa Rosa, which island we had some difficulty in finding because it was hidden in mist, while mountains on another island, which at first we had mistaken for it, were clearly visible though farther off.

There we did what was necessary. I also took the opportunity to check the oil in the engine, and discovered that instead of consuming oil the engine was apparently manufacturing it, for the level on the dipstick had risen above the maximum mark. This could have been caused by cooling water leaking in at the head gasket; but as there was no sign of emulsification in the oil, I concluded the leak must be of diesel fuel, and the most likely source of this was the lift pump which sucks fuel from the main tank to feed the injection pump. I could do nothing about it except drain and refill the sump and, as they say out there, keep my fingers and toes crossed.

Then back we sailed to spend yet another night at Cojo, where we were beginning to feel like old inhabitants. We succeeded in rounding dreaded Conception early next day, and after spending nights in two harbours, in one of which we were fog-bound for 24 hours, dropped a purse containing all our change and some vital telephone numbers irretrievably into deep water, and lost the forward ventilating cowl in a similar manner, battled on against the hard headwind to the open roadstead of San Simeon Bay. For six days we lay there, rolling heavily in the swell which made landing impossible, while the headwind blew at between 25 and 45 knots—conditions in which we can make no progress to windward either under sail or power—and gale warnings were broadcast. Another yacht was there, and her people were feeling embarrassed, for on the way up their engine had failed and they had radioed to the coastguard for assistance; but before this could arrive they had overcome their panic and effected repairs, but now were unable to get through to the coastguard to say so because the frequency was too congested with other small craft calling for help.

The wind's burst of fury was followed by a calm, and taking full advantage of this we motored the remaining 160 miles to San Francisco, where soon after dawn we sailed with a light wind in through the Golden Gate, and could see beyond the graceful red span of the bridge the buildings of the city, hazy and mysterious in the low pale sunlight. But we did not go to the city in search of a berth; instead we turned north, and having in advance been invited by the Sausalito Yacht Club to do so, tied up at its pontoon. But one night there was enough to show us that it was not a safe berth, for early in the morning fishing vessels passed by at speed, and so violent was their wash that fenders could not be kept in place; a steel cleat on the dock, to which one of our springs was made fast, broke, and the strain of the rope brought the broken piece back aboard with such force that it hit the bulwarks with a crack as though a gun had been fired. While I made that spring fast in a different place I called to mind a verse that Walt had recited to me:

> Behold the fisherman:
> He riseth early in the morning
> And disturbeth the whole household,
> For mighty are his preparations.
> He goeth forth full of hope,
> And returneth smelling of strong drink—
> And the truth is not in him.

So we moved away, and as storm warnings were now being displayed, went to the Sausalito Yacht Harbour where the manager kindly found us a berth, crowded though the place was (Plate 16).

On a small-scale chart the Bay area looks an attractive cruising ground with many well-sheltered coves and bays. But a study of the large-scale charts is disappointing, for it is then seen that the coves are not deep enough—in this connexion one has to remember that U.S. charts are reduced to mean low water, and that Bay tides frequently fall a foot or more below chart datum—while those which are deep enough are choked with moorings or marinas. These remarks do not apply so much to the Delta area, where it is possible to cruise for 100 miles inland to the city of Sacramento.

The town of Sausalito, a suburb just north of the bridge, must have been an attractive place a few years before; but we found it to be packed with hippies, tourists, and traffic; its streets were lined with shops selling curios, and its sidewalks and gutters were foul with discarded beer cans and Coke bottles. A dusty area beside the disused railway line was locally known as Basket Market or Dreams Unlimited, for there stood the rusting skeletons of several unfinished ferro-cement would-be ocean-voyaging yachts. Because this form of construction is so much cheaper than orthodox methods, there is a tendency for people to build larger yachts than perhaps they might otherwise do; then, too late, it is realized that the cost of the hull is only one-third (or less) of the total cost; money becomes tight and the project is abandoned.

However, the yacht harbour was very much to our liking because of the kind and helpful people we met there living aboard their boats. One of the most delightful of these was Connie Hitchcock who lived aboard her *Makai*. She looked and behaved as though she was in her early fifties, but rumour had it that she was over eighty. When Connie wanted a hot shower she used to fill a 100-foot length of plastic hose, plug its ends and leave it on top of the quay in the sunshine. When it was warm enough she removed one of the bungs, the water ran down by gravity, and Connie on the deck below had her shower quite unperturbed by the onlookers above. This of course could only be done on days when the fog was not thick, and in this respect Sausalito was more fortunate than most places in the Bay area, for it often remained in sunshine while a formidable sheet of fog, just as the *Pilot* said, rolled in over and under the Golden Gate bridge to smother the water and everything on it. However, although it was dense the fog layer was not always deep, and from a vantage point one could often look across the top of it and see the tips of the taller buildings in San Francisco peeping out (Plate 17 *bottom*).

Our new friends assisted us with our shopping and sight-seeing; they took us to see the redwood trees and to sample wine in the Napa Valley; they drove us over the bridge into the big white city built on hills to eat at Fisherman's Wharf, to see Nob Hill, and ride the cable cars; and when we went to pay our dues at the harbour office, we found that these had already been settled by one of our neighbours. Nicholson, too, liked Sausalito, but on getting out of my bunk one morning I noticed that he was not, as usual, curled up on Susan's bunk.

'No Nick?' I asked.

'He was here about midnight but hasn't been back since.'

When he could get ashore he always returned by dawn, and I sensed that Susan was worried.

'Let's nip ashore before too many people are about and have a look,' I suggested.

'Yes, quickly. I expect one of those big dogs on the end of this pier has chased him and put his navigation out.'

But we could not find him, and soon the place was busy with noise and bustle and we thought that our cat would hole-up until nightfall; so we returned aboard.

After dark we walked along the waterfront and down the many identical piers with yachts berthed alongside, calling softly: 'Nick. Nick?' But the only response was the groaning and squealing of the pontoons as they moved restlessly in the wash of boats passing through the outer harbour, and the teasing mewing of gulls.

'I wouldn't mind so much,' said Susan, 'if I knew he was drowned, but we'd hate going to sea without knowing. Where *could* he have got to?'

'Let's wait till the traffic eases off and try across the road.'

It was 3 a.m. before it was quiet, but our search was fruitless, and we spent another unhappy day. How does an animal remember his whereabouts when all is strange? In numerous ports in many countries Nicholson had rarely failed to come home, but on two occasions when we had shifted berth a short distance while he was ashore he had jumped aboard the wrong boat before realizing his mistake.

As I was dejectedly clearing away the evening meal Susan slipped ashore. About ten minutes later I heard her familiar step on the gangway leading down to our pontoon.

'I found him, I found him,' she cried. 'I went on to the pier parallel with this one, and he came running at my first call.'

As Susan held him, Nicholson put out a paw and pushed my chest. His purr was at full bore.

Ever since in the West Indies we came into the orbit of American yachtsmen we received much advice. These visitors nearly always showed much greater interest in our gleaming engine-room and its contents than in the gear on deck and aloft. Each appeared to have wide engineering knowledge which made me feel very humble, and always saw something that was not as it should be, and suggested improvements which I am sure were very wise; each asked me if I had such and such a spare part (something which presumably had failed in his own ship), and when I admitted that I had not, insisted that I get it.

So in time I accumulated all manner of things for few of which I ever found any call because our engine proved to be robust and reliable, and the only troubles we had were with parts not made by Ford. But I had a high regard for Walt Maertin's knowledge, and when he insisted—he could look real mean at times—that I buy myself a spare set of injectors, I obeyed him. He went further, and drove me 100 miles to the nearest Ford depot to get them. Of course we had made a note of the engine number and the part number (extracted from the Ford *Parts List*) but when on our return I offered one of the new injectors up I found it differed from the injectors our engine had, and was too big for the hole in the cylinder head by about a quarter of an inch. That problem had to be sorted out by post, but I did wonder, and I still do, why a maker cannot have a simple list of part numbers that anyone can understand. Why not start with engine model A, part No. 1? Incidentally, the part number of the injectors we needed was 2704E–9K546–A, and as a different part number was used in the U.S. for the same item, a conversion table had to be consulted, no doubt with some confusion. Hal Roth who with his wife had recently completed a cruise right round the Pacific (Plate 15*B*) in their 35-foot sloop *Whisper* (they helped look after us at Sausalito) said that whenever he bought spares for anything on board he at once fitted them to make sure that they *did* fit, and kept the original parts as spares. How wise!

Although we were not intending to make much use of the engine on the Pacific crossing, it seemed sensible to attempt to cure the leak into the lubricating oil, which we had traced to a fault in the (non-Ford) lift-pump. One of our many engine-room visitors, when told of our trouble, immediately went off and bought a complete new lift-pump together with a kit of

▶

15. Some voyaging people. *A*: Paula and Earl Schenck, who had made several trips from Hawaii to Polynesia in their immaculate *Eleuthera*. *B*: Hal and Margaret Roth, who made a bold and efficient circuit of the Pacific in their 35-foot sloop *Whisper*. *C*: Ernest Haigh with his wife Val (on his right) aboard the trimaran *Tryste II*, in which they sailed from Canada, where Ernest built her, to New Zealand and beyond with their crew of four daughters. *D*: Larry and Babe Baldwin, who had cruised extensively in their ketch *Faith*, and for many years held together the Seven Seas Cruising Association and edited its *Bulletin*.

A▲ C▼

B▲ D▼

spares for it, and made us a present of these things. But we soon discovered that the spare parts did not fit the pump, and the pump did not fit the hole in the side of the engine. However, another visitor cleverly juggled with bits from both pumps and effected a cure. I thanked him as best I could, but hoped that from now on *Wanderer* would revert to being a sailing vessel.

◀

16. Crowded though the place was, the manager of the Sausalito yacht harbour in San Francisco Bay managed to find us a berth. It was here that Nicholson went adrift for forty-eight hours.

# 5

# Hawaii and the South Pacific

On a sparkling morning in May with no sign of the fog which had been hanging around for the past week, we motored away in a calm, and with a strong, fair tide passed out through the Golden Gate, under the stately bridge across which we had so often been driven during our stay at Sausalito, bound towards Hilo some 2,000 miles away on the Island of Hawaii. We had been looking forward to this trip, for instead of the headwind against which we had been struggling for the past 3,000 miles, now the wind should be fair, and not only as far as Hawaii, but on and on down into the South Pacific, certainly as far as Fiji, and perhaps even farther; we were also looking forward to warmer, drier weather than we had been getting in the San Francisco area, and to a resumption of proper trade wind sailing.

The trip to Hawaii normally presents only one problem for a sailing vessel: her need to avoid the North-east Pacific 'high', an area of calms and light airs which tends to hover on or near the direct course between San Francisco and the islands; its position was sometimes mentioned in marine weather information broadcasts, but as the stations giving these were not very powerful (or perhaps it was the fault of our receiver) we were soon beyond their range. To avoid the 'high' it is usual to sail on a course which is the reverse of a great circle course, and this is what we did, starting by heading south; but, and this was the very last thing we expected near that coast where north winds prevail, the wind, when at last it came, was from the south; we could but hope that some north-bound yacht was making use of this unlikely happening to get up along the shore. However, this wind had one advantage for us in that *Wanderer* agreed to steer herself after we had made the precise adjustment of the mizzen sheet that she required, and for two days while the weather reports told of fog at San Francisco, we continued sailing to windward in quiet weather. Then the wind freed and freshened and the sea grew rough, and as *Wanderer* would no longer steer herself we switched on the Pinta and thereafter kept it working most of the time until Hawaii was in sight.

We passed about 115 miles seaward of our old adversary Point Conception, where an east-bound naval vessel, with water cascading over her forecastle as she plunged into the head-seas, crossed our bows probably bound for San Diego. Then, so as to pass well to the south of the 'high', we sailed for two days to the sou'-sou'-west before heading more directly for our destination. The wind moderated and came more easterly, and we ran on our way rather slowly under the mainsail and one of the twins, for the mizzen was not much use in those conditions. So chilly was the weather that we needed warm winter clothes when on deck, and for the first five days of the trip we kept the saloon heater going. This was a diesel-burning, drip-fed Kempsafe, a clean and efficient thing when properly installed with its chimney going straight up for eight feet without any bends or kinks. I had drawn it on the plans as it should be, but the Builder made the chimney with two sharp turns in it, and when I asked crossly why he had done such a silly thing, he explained that the electricians had put 10 pairs of wires behind the insulation-packed coachroof ceiling in the place where the straight-through chimney should have been. I ought to have insisted on having this altered, but, like the Builder, I had not fully appreciated how important a straight chimney is, and as a result the heater never worked as it should with the fuel vapourising and burning cleanly; though it certainly kept us warm and dry, it made a lot of oily soot which accumulated inside it and its crooked chimney, but fortunately this never reached the rigging or the deck.

Day and night the Pinta steered us faithfully but found it hard work, for as we rolled, so the rudder tended to slam from side to side, and the little electronic brain of the Pinta insisted that so long as the ship was on the chosen course the wheel must be held steady, so it had to apply through its motor a braking effect, and was therefore constantly taking power from the battery. It seemed to me, as it had earlier, that our part-balanced rudder and huge propeller aperture must be the cause of the trouble, for in all other respects *Wanderer*'s underwater parts looked like those of any normal, heavy-displacement sailing yacht. The cost of having alterations made in the U.S. would probably have been too high, but anyway our hard currency allowance was too small; however, we hoped to get something done about it in New Zealand, where the cost of skilled labour was less and (at that time) our currency unlimited.

It was wonderful to reach at last the region of the north-east trade wind, to put away our sweaters and trousers, and to feel the warmth of the sun on our bare arms and legs, and to amble on our way with ports and hatches open to the breeze which blew from aft, and to have dry decks. But the sky did lack something of the brilliance which is usual in the trades, and there was very little wildlife to be seen; there were no storm petrels, no porpoises (it was said that Japanese fishermen had killed most of them in those waters), and only a few shearwaters and bosun birds. These lovely white creatures, often called tropic birds, with the incredibly long, thin tail consisting of just two feathers, circled us at masthead height on several occasions and showed much interest in our fluttering burgee. The birds always came at breakfast time, and if Susan squawked at them they usually replied with a rasping cry. We have never seen them outside the tropics and rarely more than 500 miles from land, neither have we ever seen them dive for food. For one 48-hour period the surface of the sea was covered with what at first we took to be Portuguese men-of-war, but which proved to be velellas; these are similar but have no long, stinging tentacles. So thick were they that the sea appeared to be covered with large snowflakes; there must have been countless millions of the creatures, each with its little sail set.

The trip continued uneventfully until one day we noticed a groaning noise coming from the foredeck. At the time the stay-sail was set and sheeted amidships to ease the rolling, and each time it slammed from one side to the other the fitting on the deck to which the lower end of the forestay was secured moved

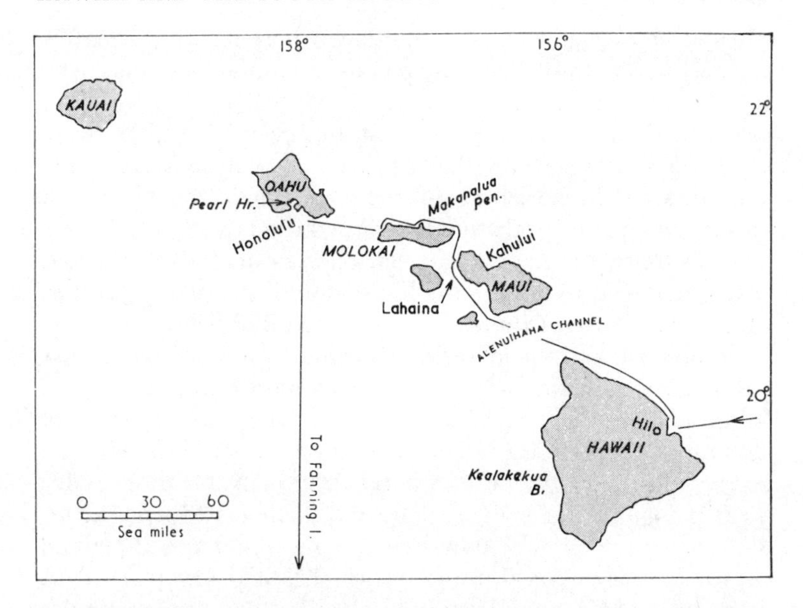


slightly in relation to the teak surrounding it. I guessed that the weld holding it to the steel deck under the teak was letting go, so we took the sail in and rove preventer lashings to two cleats to help support the stay and the mast. But later in port we discovered that the weld had not failed, but an area of the thin steel deck in way of the weld was moving like the bottom of an old-fashioned oil-can because there was no deck beam there to strengthen it. With the help of a friend we put eyebolts each side of the fitting, a bolt through all three, and a strengthening plate under the deck, and that fixed it.

As we approached Hawaii a navigational problem presented itself. The sun, increasing its northerly declination, was daily moving north to meet us while we were creeping south, so that on the day when the sun passed overhead I could not obtain a noon sight for latitude. It might be thought that this would not matter much, for one has only to cross an afternoon position line with one brought forward from the forenoon to obtain a fix. But on an occasion such as this the sun's azimuth (true bearing) all forenoon is nearly east and all afternoon nearly west, so the position lines cannot cross. We had been in similar situations on other voyages, but always in positions where it did not matter as we were yet many days from making a landfall, and had been able to get a fix from the stars if we wished to. On this trip, however, the sky was often overcast, but even when there were no clouds it was so hazy that the stars were not visible at the same time as the horizon could be seen. But in the evening of the day when at noon the sun had passed right overhead, casting shadows that were grotesquely shortened to small dark patches, Susan managed to obtain an observation of the planet Jupiter, which was so bright that it was clearly visible before any stars could be seen, and the following morning I was able to cross her position line brought forward with one obtained from the moon. In an article in a magazine I had mentioned the difficulty of obtaining a fix in the above conditions, and both Michael Richey, secretary of the Royal Institute of Navigation and a great navigator in his own right, and Nicholas Morris, Lieut. R.N., were kind enough to write to tell me about an old method known as 'very high altitudes' of which I had never heard. A description of this would be out of place here, but the method is described both in Lecky's *Wrinkles in Practical Navigation* and Burton's *A Manual of Modern Navigation.*

The port of Hilo, protected by a low breakwater, lies on the windward side of Hawaii, which is the largest of the Hawaiian Islands and is generally known as the Big Island to avoid confusion with the State of Hawaii (the fiftieth state of the U.S.A.), which includes all the islands of the group. We came to it in the afternoon of our 20th day at sea, and were met purely by chance by John Lavery (he with his wife Mary had sailed round the world in *Si-ti-si* some years before and passed us in *Wanderer III* in the Red Sea). He had been racing, and had with him in his sailing dinghy the captain of the coastguard cutter; they led us into the inner harbour, showed us where to anchor and, still under sail, ran our sternlines out to the wall. There on one side of us lay the smart coastguard cutter, white with a

▶

17. Steamboat a-comin'. Ray Quint, our 'tour director', photographs the stern-wheeler *Mark Twain* in what must surely be the best of all entertainment parks—California's Disneyland. *Bottom*: Fog rolls in through the Golden Gate. Though it is dense it is not very deep, and from a vantage point one can see the tips of San Francisco's taller buildings peeping out above it.

diagonal red stripe on her topsides—these cutters, which often look like miniature destroyers, do in U.S. waters (among other things) what in the U.K. is done by boats of the Royal National Lifeboat Institution—and on the other side a fine stand of casuarinas. We had just made all snug and spread our awning when we were told that a seismic sea wave warning had been issued; there had just been an earthquake in Peru causing the loss of 50,000 lives, and ever since the disastrous wave of 1 April 1946 Hawaii, which suffered damage estimated at $25 million, was highly conscious of the danger attending such a phenomenon. We were told that we and the coastguard cutter must make ourselves ready to put to sea, which in such circumstances is the only safe thing to do; but much to our relief the all-clear was given soon after.

Our berth was quiet, and comfortable once we had rigged the mosquito screens in our sleeping cabin, but with no public transport was inconveniently far from the town; also there was plenty of rain, for Hilo gets 140 inches in a year. But the Laverys, though working hard to make money enough for the building of a new *Si-ti-si* (as I write this she is taking shape in Tasmania) always found time to drive us wherever we wished to go, and we were hospitably entertained by them or others nearly every evening of our stay. Sugar cane and volcanoes were the big things in that high and unspoilt island with its vast airy spaces sloping towards the sea, and tourism had only a tiny foothold, the hotels being sequestered near one beach. We were fortunate to see, in company with John and Mary, a volcano erupting, the red-hot lava boiling out of its crater to flow for miles down the mountainside carrying all before it.

The Hawaiian Islands lie within the trade wind belt, and their high mountains cause the wind to funnel through the channels between with increased strength, often reaching gale force, and raising a rough sea. In my rather limited experience

◀

18. *Top*: We anchored for a night off the peninsula on Molokai's otherwise uninhabited north shore, and found it hard to believe that this neat little village with its hospital, church, and dwellings, had only 100 years before been the hellish lazaretto to which lepers from Oahu were transported to spend their last days in squalor and neglect. *Bottom*: At Honolulu high-rise hotels and apartment blocks dominate the famous Waikiki Beach and the Ala Wai Yacht Harbour.

they did not seem to offer a very good cruising ground, attractive thought it undoubtedly is, for sheltered harbours are few, and, apart from Hilo, were overfilled with yachts, as indeed are most places in U.S. territory. Open anchorages on the leeward sides of the islands are considered to be safe, except in the winter when south-west gales spring up with little warning, but they usually have onshire winds during the day together with some swell, and all too often the bottom is of coral. On his third voyage, Cook in the *Resolution*, with the consort *Discovery*, spent eight weeks crusing among the islands looking for a harbour in which to refit, but without success, for the now famous Pearl Harbor, in which is the big U.S. navy base where the Japanese launched their devastating air attack in 1941, only became available after channels had been dredged and blasted through the reefs. Eventually, having failed to find anything better, Cook anchored in Kealakakua Bay, which is on the west coast of the Big Island and is wide open to the west and south. He remained for two months, refitting and refreshing, but soon after putting to sea in search of the North-West Passage, *Resolution*, due to incredibly bad workmanship and materials provided while in an English dockyard, sprung her lower foremast; this was a serious disablement enabling her to set only seven of her twelve sails, and to effect repairs in the only reasonably smooth water known to him, Cook put back to Kealakakua Bay where, owing to an entirely unnecessary misunderstanding on the part of the priests, who originally had believed he was some sort of a god, he was killed with clubs and daggers on the beach.

No doubt it was unenterprising of us not to visit that sad and historic spot, but when we left Hilo with the coastguard cutter escorting us out to sea, the weather was not suitable for going there, as the wind was ahead and light, and the sea so steep and confused that we could make no progress in that direction; so instead we turned to the north-west, and ran slowly along the coast. In the evening the wind freshened, and as by night we tore through the notorious Alenuihaha Channel, I was careless enough to break one of my little toes while reefing the mainsail, which by then was the only sail we still had set. It was a particularly dark and wild night, and when the light on the south point of the Island of Maui, for which we had been heading by compass, did not appear at the expected time—and we did not know what its range was, for this was given variously as 7, 10 and 14 miles on the three charts we had covering the area—we hove-to and waited for dawn, which because of the overcast sky was late in coming, and then continued on our way.

In a windless pocket on the lee side of Maui, into which we had motored, we stopped outside the small artificial harbour of Lahaina. Within its little rough-rock breakwater it was only 800 feet long and 200 feet wide; from the offing it appeared to be full almost to overflowing, an impression heightened by the fact that one yacht was lying at anchor outside rolling in the swell.

'What we need,' I said to Susan as we were reluctantly turning away, 'is some nice young man to come out and tell us there is room, and offer to lend us a hand.' I had barely spoken when a dinghy appeared from behind the breakwater and came rowing purposefully towards us. The bearded young man in her said he had recognized *Wanderer* from a photograph he had seen. 'There's a berth right next to my ketch,' he said, 'and if you'll drop your hook over at the far side of the harbour, I'll take your sternlines and make them fast on the breakwater.'

So in we cautiously went. The place was indeed very small and very crowded, but soon we were moored close to Bill Sellers's *Moonglow*. Bill, an artist, and his wife took us to see something of the attractive island, and they told us about the yacht bound for Hilo from Sausalito. No sights were obtained in the final few days, but she made port—'The marks were right'—and the owner turned in while his crew went ashore. Three hours later they returned, and one of them said: 'Skipper, this isn't Hilo, it's Kahului.' Kahului is on the north coast of Maui and more than 100 miles from Hilo.

The usual route from Maui to Honolulu (capital of the group on the island of Oahu) passes south of the island of Molokai, along the coast of which are several windy reef anchorages. But Bill had recommended us to go north of Molokai because of the scenery. We were glad we took his advice, for the eastern half of Molokai's north coast was one of the most exciting things we had ever seen from seaward. Cliffs rose almost sheer from the breakers for 2,000 and 3,000 feet, and down their rugged faces streaked the silver threads of many waterfalls, some with an uninterrupted leap of 500 feet. Now and again the cliffs turned sharply inland to allow a glimpse of some dark and inaccessible valley, perhaps with a stray shaft of sunlight picking out a vivid patch of verdure. This high rampart pulled the rain out of the trade wind clouds, and although we were sailing mostly in sunshine, the land under the cloud mass was gloomy, frequently shrouded in rain, so that each moment presented a different and often forbidding aspect. Until we reached the leper settlement on the low Makanalua peninsula (Plate 18, *top*) half way along that fantastic coast, we saw no sign of man. Inshore the waters had not been surveyed, and on the chart a wide band along the coast was devoid of soundings, so we did not approach as closely as we would have liked to do from a photographic point of view.

We spent a night at anchor on a rocky bottom in the lee of the peninsula, where a white-robed nun was fishing from a reef, and found it hard to realize that where the leper settlement with its tidy, modern dwellings, its little white church and hospital, and its airstrip, now stood, had, only 100 years ago, been the hellish lazaretto without buildings or facilities of any kind to which lepers from Oahu were deported to die neglected in squalor and misery. We were not invited to land, and as we did not have the required permit, we did not attempt to do so. The following morning we sailed to the rainless west end of the island, and spent a very windy and restless night at anchor there before continuing to the big, modern city of Honolulu.

Entering the Ala Wai yacht harbour, where it is said if you want a dock your name will probably be on the waiting list for more than a year, we were given a berth for two weeks at the Hawaii Yacht Club, with bow lines to the shore and a rather vulnerable, because of traffic, anchor on a warp out astern. All around us lay yachts of every size and sort packed like sardines in a tin; some of them suggested that they had been there for a very long time and were never likely to return to the mainland

▶

19. *Top*: On the way to Fanning Island we passed through the doldrum belt with its squalls and calms and variable currents. *Bottom*: But on reaching the South Pacific we had a steady beam wind and sometimes carried our maximum spread of sail. The glamorous mizzen staysail steals the picture, but also to be seen is the radar reflector which we carry permanently lashed to the main backstay.

NESTLE
NESCAFE
NESCAFE

from which they came, for it is one thing to run downwind to Hawaii and quite another to beat back. But among the throng were some fine, able-looking craft whose people were friendly and helpful, and it was particularly good to meet Earl and Paula Schenck (Plate 15*A*) in their lovely steel ketch *Eleuthera* (her bilges were painted with white enamel), for we had by coconut radio heard much about them and their several voyages to the Marquesas.

I always find my first drive through a big town exciting, the more so if the visit is by night, and it was Earl and Paula who drove us in a big limousine through the brilliantly lighted streets of Honolulu, which seemed to be filled with a great tide of swirling traffic; skyscrapers towered on either side, their tiers of lighted windows and balconies soaring up into the black velvet sky; tantalizing smells and snatches of music came from open doorways, and there was an air of easy-going gaiety about the crowded sidewalks. The Schencks drove us out to dine in the house near Diamond Head in which they were temporarily living while they repainted *Eleuthera*'s insides; with its patio bathing pool under-water-lighted, its array of electric gadgets in the kitchen, and 'contoured' seats in the water-closets, it might have come straight out of some glossy magazine, or perhaps from the set of a Hollywood studio.

I sometimes wondered just what it could be that brought so many holidaymakers (more than a million a year) to Honolulu to stay in the honeycombs of the vast, high-rise hotels (Plate 18, *bottom*), many of which crowded the famous Waikiki Beach. In a line, shoulder to shoulder, each vied with its neighbours to dominate the scene, so that many of the occupants looking out of their expensive apartment windows could see little but the grey concrete ramparts of the neighbouring monsters. It could scarcely have been the wide, golden beach that was the attraction, for few people made use of that, the hotel guests preferring to do their sunbathing in more sophisticated surroundings. As

◀

20. *Top*: Phil Palmer and John Fleetwood of Fanning Island were surprised and delighted when we arrived with some of their long overdue cargo, even though the case of rum was missing, and they piloted us, *bottom*, to a good anchorage behind Cartwright Point in a setting that suited *Wanderer* so well.

we strolled along the beach in front of the hotels, we could look in through railings and see pink bodies stretched out on chaise-longues crowded round the turquoise pools, with white-coated waiters bringing trays of iced drinks. Island music dripped out of loudspeakers, and there was a smell of hot flesh and sun-lotion on the stagnant air. More high-rise blocks were being built, and all through the day there was the nagging beat of piledrivers, while miniature sandstorms swirled round the building lots. Dense motor traffic snarled and tangled in the streets, and every few minutes during the morning and evening rush hours, observers in helicopters broadcast the traffic situation and reported, almost with pride, the latest accidents. Meanwhile the jumbo jets roared in with more and still more visitors. As a passenger in one of these you could pay, if you wished, a little extra for your ticket so that on arrival at the airport a pretty girl would garland you with leis; but instead of the cool, fragrant flowers of Polynesian tradition, the leis would be of plastic.

When it became known on the waterfront that we would shortly be leaving bound towards Fanning Island, 1,000 miles to the south, we were asked to take along some mail and cargo. This, which comprised five large cartons and some small packets, a nylon fishing net, a football, and an enormous kettle, had earlier been entrusted to a large power yacht at Christmas Island, but due to overcast weather and wayward currents, she had failed to find Fanning and had brought the cargo on to Honolulu. We stowed it in the engine-room, which until then I had always thought was much too large, and as the yacht club bell rang in farewell (that is a gesture which I believe is always made there when a yacht leaves on a long trip) we departed from the Ala Wai harbour as planned on the last day of June.

Our course took us 100 miles to leeward of Hawaii, but even at that distance the Big Island, which rises to more than 13,000 feet, and so is a little taller than Tenerife in the Canaries, interfered with the weather. We sailed in under a pall of low, black cloud which discharged torrents of rain on us, there were light variable winds and squalls from all directions, and eventually a calm with a steep, confused sea. After sailing about 140 miles we had had enough of this and decided to motor out from under and look for better weather, but after running for a few minutes the engine lost power and made a lot of smoke. We could expect to have plenty of wind all the way to Fiji, except for the crossing of the doldrum belt, and any sailing vessel can manage that, so the failure of the engine would not have mattered in the least except that we had promised to deliver our cargo at Fanning; although the pass into the lagoon there is straight and easy, we knew it to be narrow and to windward, and felt that a reliable engine would probably be needed to get us through it. The motion was so wild that I could do little except check the fuel filters and pumps, which seemed to be in order, so we turned round and sailed all the way back to Ala Wai harbour and its envelope of cruising yacht friendship, the seal of which we had broken with difficulty and regret only three days before. There we were guests of friends for dinner, and next day with the help of a young engineer set about investigating the trouble. It turned out that part of the cone of one of the fuel injectors had broken off, and we could but hope that piece of hardened steel had gone out of the exhaust and was not still sculling about inside the engine. We fitted the spare set of injectors which Walt Maertins had so wisely insisted that I buy, and 48 hours after our return got ready to leave again.

In our simple little 30-footer *Wanderer III* we reckoned we could prepare for sea from a harbour stow in less than 15 minutes. But in our more sophisticated 49-foot *Wanderer IV* the operation takes us about two hours, and if you should wonder what this entails and how we get our exercise, here is a list of the things we have to do:

Close the sea-cocks of heads and basins (6); remove mosquito screens and close all ports (13); bolt all locker doors (27) as their spring latches are not seaworthy; extinguish the refrigerator, empty its parafin tank and close its deck vent; ship the galley and table fiddles and the non-slip mats; stow contents of shelves in the after heads compartment and unrig shower hose; unrig and stow cockpit light; stow cockpit cushions; close gangway and stow boarding ladder and fenders; lace on weathercloths each side of cockpit; furl and stow two awnings and windsail; lash dinghy oars, hoist dinghy, capsize and secure it;

remove and stow sailcoats; unbag and hank on two headsails; take all halyards from their idle (non-flap) positions to the masts, and shackle to heads of sails; reeve and shackle on two pairs of headsail sheets; uncoat binnacle and wheel. The reason for having so many coats and covers and for keeping so many things below is to protect them from the sun, which in low latitudes is highly destructive of sails and gear. If we have been lying alongside, the lines, fenders and chafing gear will have to be stowed after we have cast off; but if we have been lying to an anchor, after weighing, the anchor will have to be lashed, the navel pipe plugged, the windlass coated, and the hose used for washing the cable and anchor uncoupled and stowed away.

In silence Earl, who is a perfectionist, watched us do these things, and we became so nervous under his eye that some of them we did not do properly. Then he let our lines go, we backed away from the dock and turned, the club again rang its bell, and friends waved from the lawn. We made sail in the harbour and hurried out through the entrance channel, which is about 600 yards long and about 70 yards wide. Its western side is bounded by Magic Island, on which stood the tents and wheels and coasters of a fair, and its eastern side by a reef on which the swell (it was higher than usual that day) broke heavily; on it bronzed figures were surfing only a short distance from us. At sea the fresh wind was just abaft the beam, and under all plain sail we had been moving fast for about an hour when Susan, who was steering, said those familiar and dreaded words: 'There's something wrong with the steering. I can hardly turn the wheel.'

At least I knew that this time the grease-gun, which had gone adrift off Point Conception, was properly stowed, and in the engine-room and afterpeak I could see nothing wrong. But each time Susan forced the wheel over a spoke or two there was a terrible rending noise as though a sheet of plywood was being ripped. I could not trace that noise; wherever I listened in the steel hull it seemed to be coming from some other place.

Somehow we got the ship turned round and headed back towards the harbour, and now almost close-hauled she steered herself with no call on the steering gear. Meanwhile we moved the dinghy out of the way, unscrewed the plate in the deck over the rudder, and tried to ship the steel auxiliary tiller, only to find that the square head of the rudder stock refused to accept it. Then, too late, I remembered that when we had steering trouble in Mexico I had used Walt's heavy hammer on it and presumably had distorted it, and probably many hours of work with a file would be needed to restore it to its original shape, for only the tip of a file could be got into the small space available. We discussed anchoring outside the harbour until someone should come to help us in, but without being able to steer properly it was questionable whether we could be towed safely through the channel; as I have said, the swell was heavy that day, and no vessel could secure alongside without risk of inflicting damage on herself or on us. So with more dreadful rending noises Susan got *Wanderer* lined up with the entrance channel so that she might go through it without having to be steered, for we feared that at any moment a further movement of the wheel might break some vital part of the gear and put us on the reef. With fast-beating hearts we passed safely through into smooth water. Fortunately the berth we had so recently vacated was still empty, and at our hail people came running to take our lines and fend us off, and Earl boarded us almost at once with offers of assistance.

Where the steering gear shafts pass through each of three bulkheads on the way from the cockpit to the stern they are supported by ball-races, and investigation showed that one of these, where the shaft enters the afterpeak, had disintegrated—traces of yellow-metal and some broken balls lay beneath it. The race was provided with a nipple, and I had been meticulous about injecting grease, but as the afterpeak was reached by that yachting absurdity a flush-fitting hatch, rain or spray could enter easily when the ship was heeled, for the waterways provided did not then collect the drips. It was so wet a place in bad weather that I had rigged up a small pump to keep it clear; nevertheless water had, no doubt, reached the ball-race at one time or another. The local boatyard sent along a good engineer to fit a new one, and he reported that the broken race was not of the self-aligning type (as it should have been) and that this, not water, had caused the failure. Again we were guests at dinner, but we asked the club not to ring its bell this time, and when we left two days later it was, to quote R. T. McMullen, 'not without a taste of brimstone in the mouth' as we wondered what could possibly go wrong next.

Short of rebuilding the afterpeak hatch I could think of no satisfactory way of making it watertight, but Susan hit on a simple solution; she suggested we seal it with paint-masking tape. I thought the water would soon lift this or wash it off, but I was wrong, and we found that provided it was applied to a dry surface it lasted right through a long passage and at the end had to be removed with a scraper. Naturally the tape was inconvenient, and to make the hatch properly watertight was one more of the jobs to be put down on the New Zealand list.

On this occasion the Big Island let us pass without much trouble, and on the third day out we were sailing fast with a strong wind on the beam, a point of sailing on which *Wanderer*, like most other yachts, is difficult to steer, and the Pinta was making heavy weather of it. I felt that somehow it must be possible to get the ship to steer herself, but even with the mainsheet eased off until the sail was lifting, she still persisted in coming up a long way to windward of the course. Oddly enough handing the jib improved matters, and after we had rolled a deep reef in the mainsail we got her to jog along only 10° off course and gave the Pinta and ourselves a rest.

For several days we had the strong beam wind and a clear though hazy sky, and after a good sleep we made more sail to take advantage of it; although we then had to steer it was such grand sailing that we did not mind, and soon the whole ship became caked with salt as the spray driving over her was repeatedly dried by the sun. At about that time we noticed the paint cracking round the bases of the halyard winches, of which there are three on the main mast, but not until reaching Fanning did we investigate this, when we found that each was held by four small brass screws and had no bedding compound, nor even any paint, behind it. There we bedded the winches on Bostik and secured them with bronze screws of a proper size.

In latitude 7°N., rather earlier than we expected, we came into the rains, calms, and occasional though never violent squalls of the doldrums which lie between the two trade wind belts (Plate 19, *top*), and used the engine to help us along a little.

Fanning, which is one of the Line Islands and lies in about 4°N., is an atoll only a few feet high, and the 70- to 90-foot palms growing on it may be sighted from a small craft when 10 or 15 miles off. It is not always an easy place for the navigator coming from the north to find. On his way down from Hawaii he will be in the North Equatorial Current, which may set west or north-west at up to 18 miles a day. Then he has to cross the Equatorial Counter Current, which in July can do up to 30 miles a day mostly in an easterly direction. This is about 200 miles wide, and Fanning lies just beyond it in the South Equatorial Current, which can do 24 miles a day in a west or north-west direction. There is no indication of the boundaries of these currents, except possibly a profusion of fish, but if one could rely on the above directions and figures, which are taken from the July pilot chart, there would be no problem; however, one can of course not rely on currents to do what man predicts.

In circumstances such as these it is often too long to wait to get a fix by crossing two position lines obtained from observations of the sun, and it is then that star sights are so valuable. But, as we had noticed in California and on the trip to Hilo, though the sky might be cloudless the stars did not shine with their accustomed brilliance, and observations of them were rarely possible. In saying this I am well aware of the fact that as one grows older the telephone directories are printed in smaller and smaller type; my sight is certainly failing, but Susan's is excellent, and it was she who took the star sights when these were at all possible. It is understood that in low latitudes stars do not shine so brightly as they do in high latitudes; nevertheless it did seem to us that the sky was veiled with a thin film of vapour; by day this might not have been noticeable except that never in the North Pacific did I need to use the darkest sextant shade for sun observations. In California I knew the haze to be the locally produced smog, which in Los Angeles was causing such concern that there was talk of banning cars—one has only to fly over that city sprawled in its windless, mountain- and sea-girt basin to appreciate the density of this man-made pollution. While we were at Honolulu an American who had lived there for many years took us for a drive to the Punchbowl, and while we were admiring the view he remarked that from there in the past one could always see Molokai (except in rain) but that during the past few years it had rarely been visible. This he put down to smog.

However this may be, the fact remained that when in the North Pacific we needed stars they were rarely bright enough to be observed, and so it was now. For several days I was unable to get a noon latitude because of cloud, but the planets Venus and Jupiter were sometimes visible when needed, and at dawn of our tenth day at sea Susan did manage to get an observation of Rigel (one of Orion's feet), which gave us a useful position line showing that we were too far to the east. We altered course accordingly, and a couple of hours later as we lifted on the swell a line of green palm fronds rose for a moment above the horizon across our bows—our first glimpse of Fanning, and what a rewarding moment such a landfall is! Under power now in absolute calm we slowly raised the island—we had sighted the tops of the palms at a distance of 14 miles—passed the disused cable station, rounded the breakers off Danger Point, slipped in through the pass at low water but with a strong stream against us, and before noon were at anchor in the vivid lagoon.

Fanning is owned by Burns Philp, the big Pacific trading company, and is under Gilbert and Ellice government. It is an excellent example of an atoll, for its coral reef, which encloses a lagoon 9 miles long by 4 miles wide, is almost completely covered with palm-clad motus, there being only one deep pass and a few shallow canoe passages into the lagoon. Because of coral patches not much of the lagoon is navigable except by canoe, and we brought up in the recognized anchorage off the settlement just within the pass and on the eastern side of it. But there the tidal streams ran hard, and very soon Phil Palmer, who had been B.P.'s manager on the island for 30 years, and John Fleetwood, his New Zealand assistant (these were the only white people there) came off to pilot us to a better anchorage the other side of the pass in behind Cartwright Point (Plate 20, *bottom*), where we found a depth of 10 feet between the coral heads, and no tidal stream; but there was not swinging room, so after we had anchored we ran a line to the shore.

As only two ships a year called at the island with supplies and to load copra, which is the only export, and as the transmitter in Phil's house was powerful enough only to talk with neighbouring Washington Island, Phil of course had not heard of us and was surprised and delighted when we told him we had the things he wanted from Christmas Island and mail from Honolulu (Plate 20, *top*).

'Bless you,' he said, 'I can do with that case of rum. I'm nearly out.'

I had to tell him at once that although we had not opened the cartons I felt sure that none of them contained spirits; and so it proved. Among the mail was a letter from the people who had originally attempted to bring the cargo (including rum) from Christmas Island, and it explained that the rum had been jettisoned at sea to avoid complications with customs at Honolulu. Phil took this blow cheerfully, for he was accustomed to having things for which he had paid vanish on the way to his island; he gave us the key to a 500-gallon rainwater tank on the shore close by, and that evening sent a launch over with excellent fresh bread for us and fish for Nicholson.

Phil was married to an Australian who had lived with him on Fanning for 20 years, but by then their two daughters had reached an age at which their parents thought they would be better away from the island, and anyway they needed educating. So Mrs. Palmer took them to Australia, and although she and her husband were on good terms and corresponded with one another, she never returned to Fanning. John married an attractive Gilbertese and was the father of several children.

Situated at Cartwright Point we were cut off from the settlement only a quarter of a mile away by the 5-knot stream in the pass (Plate 21); this made it inadvisable to attempt the crossing by dinghy; indeed, it was not uncommon for canoes to be swept out to sea on the ebb, and some islanders had been lost in that manner. However, nearly every day of our two-week stay Phil or John came to visit us, or to take us across for a few hours at their homes, or to drive us about the island, or the Chinaman who looked after the disused cable station on our side would take us across in his boat to the settlement (Plate 22) where many of the 400 Gilbertese lived, and where there were two churches, two stores, a meeting-house, and a policeman. So we did not feel lonely or isolated, but after the excitements of new friends and the bustling waterfronts of California and Hawaii, it was restful to know that no strange voice would hail: 'Aboard the *Wanderer*', and we were able to get on with the business of maintenance, writing and photography, bathing in the pale green translucent water—though not in the morning or evening when a sting-ray of great size majestically cruised in our bay—and just plain enjoyment of our handsome ship lying in the colourful lagoon with its frieze of graceful palms, a setting which suited her so well. Nicholson enjoyed expeditions ashore, though he treated the numerous land-crabs and brown men on bicycles with suspicion, and by night watched fascinated as great shoals of phosphorescent fish progressed in leapfrog fashion round us.

To earn his basic pay each working man on the island had to make 300 lb (3 sacks) of copra a day. The Bank Line ship which called twice a year to carry the copra to the U.K. could not enter the lagoon, which is too shallow, or anchor outside, where it is too deep, so she lay-to outside and the copra was taken to her in double-ended surf-boats towed by diesel-driven launches. After Fanning's copra had been loaded, the boats and launches, together with their crews, were put aboard and taken to Washington Island, 80 miles away. That island has no lagoon and no good landing, but an 11-inch cable runs from the beach out to a buoy, and after the surf-boats had been launched from the ship, each in turn was towed in close to the cable where two of her crew of 7 men went overboard and hitched lines from bow and stern to the cable, which was then hove up and dropped into fairleads at bow and stern, and an eccentric brake engaged it. As the boat was carried in towards the shore on a swell she slid along the cable, but when the backwash started the brake was engaged to hold her there until the next swell came along to carry her further inshore. After the sacks of copra had been loaded into her the process was reversed, the backwash being used to carry her out from the shore; the cable was hove overboard and the launch towed her alongside the ship where the

▶

21. From the masthead we could look over the point and across the pass, where dark blue water shows it to be deep, towards the settlement a quarter of a mile away . . .

sacks of copra were hoisted aboard. Constant care and first-class seamanship is required to carry out such a procedure again and again without disaster, and of course to manoeuvre the ship near enough to the shore; but a few years earlier the *South Bank* did go ashore on the reef at Washington and became a total loss. Her second mate was killed when a boat he was in was lifted by the swell up and under a sling of cargo hanging over the side of the wrecked ship, and he was buried ashore. But while we were at Fanning there was a great to-do as the dead man's parents were suing the Bank Line. On instructions from the U.K. his remains had been exhumed and sent home in the last ship to call, but a thigh bone was missing. The questions Phil, John and the manager at Washington were asking themselves were: was it Gilbertese sorcery or vandalism, and would anyone notice?

All too soon our highly enjoyable time at Fanning was up. We left on the first day of August, carrying the mail, and Phil and John came over by launch to pilot us out of our lovely, coral-speckled berth. It was with real regret that we said goodbye to these two fine, independent men, who after 11 years together still seemed to enjoy each other's company, and who so loved their island and the life on it that they found their periods of enforced leave dull and frustrating. We headed for Apia in Western Samoa, 1,300 miles to the south-west. Soon the palms sank into the sea astern, and once again we were out alone on the wide and empty ocean. In this connection it may be of interest to note that between Hawaii and Fanning we saw one warship (among the islands) and one fishing vessel later; between Fanning and Samoa we saw no ships at all, and between Samoa and Fiji only one, hull down.

The trip on which we were now embarked proved to be an easy one. There were no navigational problems, and the wind was on the beam all the way. This is the condition in which the ketch, an indifferent performer on other points of sailing, should prove her worth; but because of the difficulty of steering

◀

22. *Top*: . . . where we found the Gilbertese busy making copra, the island's only export, and, *bottom*, the children very interested in whatever it was Susan was trying to tell them.

*Wanderer*, we could only carry the 200-square-foot mizzen on days when the wind was moderate, for when the wind was strong this sail caused too much weather helm. Partly as a result of this we never made a day's run in excess of 157 miles, and with our 40-foot waterline we ought to have done better than that. *Wanderer III*, with her 26-foot waterline, was certainly no flyer, but she often did as well as this and once achieved a day's run of 169 miles. Of course we were not a powerful enough crew to drive our big, heavy-displacement ship as hard as she needs to be driven if she is to make fast passages, but we had high hopes of improving her performance with the alterations we intended making in New Zealand. However, in the lighter winds that prevailed at times we were able to set our grand total of 1,640 square feet of sail—mainsail, staysail and mizzen, and our two light-weather sails, genoa and mizzen staysail (Plate 19, *bottom*). This was always interesting and certainly looked impressive, but we did sometimes wonder how easy it would be to reduce sail quickly in the event of a sudden squall.

On most days we had a noticeable set to the westward, just as one would expect in those latitudes. Our practice was to steer by hand through the night and part of the forenoon to save the Pinta, but to let the latter take over and grind away through the blazing afternoons which were much too hot to be enjoyable. Even below we found it too hot, for with the wind on the beam our normally efficient ventilation by way of the hatches failed, and we realized that we needed more and larger cowls.

It must surely have been by chance, but as we left the equator astern the stars became noticeably brighter, and they, as so often in the past, enlivened the night watches. Early in the night at the start of this passage, Venus and the new moon were having a race down to the west horizon. Astern the handle of the Plough curved up to point out Arcturus (very high) and Spica, with the near-by gaff sail pointing its peak at her. Jupiter was mostly hidden behind our mainsail, but fine on the port bow and useful to steer by in the early part of the night lay the Southern Cross, rather on its side, and with its pointers apparently about to pounce on it. In the middle watch there was little I could recognize, but before dawn Orion came striding up over the eastern horizon closely followed by Sirius and Canopus. By then the sky was beginning to pale, and all too soon the sun came lurching up to scorch one's eyeballs, and it was time to stretch, switch on the Pinta and get breakfast.

So our nights and our days passed pleasantly enough, except for the heat, and uneventfully, except for one small excitement: some years before I had attended, though I was not a part of it, a sort of brains trust at the Royal Lymington Yacht Club. One of the questions asked was 'What would you do if your mainsheet carried away at the boom-end?' The panel had no very good solution to this problem, and I had often wondered what I would do in that situation, especially if there was much wind and sea, and failed to provide an answer. Well, our mainsheet did let go.

As we have roller reefing gear the end of the boom is provided with a swivel fitting to take the sheet and topping lift, but the firm that made this so arranged things that as the boom turns in the act of reefing, the lug to which the clew of the sail is outhauled fouls the eye of the pin of the shackle which secures the mainsheet block to the swivel fitting. The obvious cure for this is to reverse the shackle so that the eye of its pin is aft, but unfortunately the designer of the yacht had allowed so little space between the boom-end and the mizzen forward shrouds that the eye then fouled the shrouds when tacking or gybing. I wanted to cut a bit off the boom, but had not done so because that might have called for an alteration to the leech of the sail, so instead I had replaced the usual shackle with one in which the pin had a screwdriver slot instead of an eye. Although I had done this up as tightly as possible, and often checked it with a screwdriver, it had, unnoticed, become unscrewed, the ½-inch diameter shackle had opened up like a bit of wire and dropped the block with the sheet on deck. Fortunately for us the boom-guy was in use at the time, and with this we were able to haul the boom inboard and fit a new shackle, but this time the shackle was of the kind that fouled the lug, so we had to be very careful when reefing. Incidentally, I was astonished at the wear and tear our big ship imposed on her various fittings. For example, the mainsheet and topping lift shackles ground away the yellow-metal bails to which they were attached, and the shackle at the clew did a similar thing, so I had to build up the weakened parts with Marinetex to stop further wear, but found this did not last for long.

That evening we drank our last drop of scotch—there had been a bit of a run on this at Fanning—and soon after we were struck by a wind and rain squall of great violence. As the mizzen was set at the time the pull on the helm was great, but I managed to force it up and bear away while we hastily reduced sail. Clearly this had not been one of our better days, and as the rain soon killed the wind to leave us slamming helplessly in the deluge, we took all sail in, hung up a light, and both turned in for a much needed full night's sleep.

Soon after we had finished breakfast on our 11th day at sea we could make out ahead the mountains of Upolu, the Western Samoan island for which we were bound. The harbour of Apia, which lies on the island's north coast, is protected in some measure by reefs each side of it, but on the chart looks to be a rather open place for small craft. We picked out the red-and-white leading beacons a long way off before the buildings in the town were visible, and passed in between the surf-enveloped reefs to find the harbour remarkably free from swell.

After the officials had finished with us at the quay, I asked the harbourmaster to suggest an anchorage, and he pointed out an area off the river mouth where, he said, the holding was excellent. I wondered about this, for the *Pilot* states that the holding ground—a thin layer of silt over flat rock—is bad, and not until the third attempt did we find a spot where the anchor showed any sign of taking hold. There were two other yachts in the port—one of them was Lawson Burrows's trimaran *Roulette*—and we soon learnt from their owners that they had both dragged; one had been saved in the nick of time when only a few yards from the reef, the other had gone on the reef and lost her rudder; one might have supposed such happenings would be known to the harbourmaster. Eventually we tied up to an enormous, disused mooring buoy, but could not leave the yacht unattended because of the risk of her bumping into it during the frequent calms.

However, there were other reasons why during our week there I did not care to go to the shore except to ferry Susan to and fro on her shopping expeditions or to get water, and this was a mistrust of the Samoans. Apia has always had a bad reputation for theft from yachts, and I did not care for the fact that several people invited themselves on board, and all wanted to know how many people there were in our crew; we also had an unpleasant encounter with a fat man who called himself Jack and demanded money with threats. When over an evening drink aboard *Roulette* I mentioned this to Lawson, who knew the Samoas and other Pacific islands very well, he remarked:

'It's a sad thing, but I have found the more churches a place has the more dishonest are the people, and just look at the town here'; it fairly bristled with ecclesiastical towers and spires. 'Tell you something else,' he added; 'if you leave your dinghy on the beach the kiddies will push it off just for the fun of watching it smash up on that reef down wind.'

One night when the inter-island ship *Tofua* was expected, the light offshore breeze had swung us with our stern out in the deep water, so we had hung the riding light right aft where it illuminated the stern deck above our sleeping cabin. Susan and I had been deeply asleep for some hours when I was awakened by something repeatedly slapping my naked torso. Sleepily I looked up and saw in the radiance of the riding light that a girl wearing nothing but a pair of thin briefs was squatting by the edge of the open skylight and hitting me with my shirt, which she must have fetched from the saloon where I had left it, for the night was hot. Susan awoke at the same moment and sat up in her bunk.

'What are you doing here?' she asked, shocked, as I was, that someone could have got aboard without waking us in the process.

'Please help me,' said the girl in good English, 'my father is trying to murder me. I swam out.'

She was a pretty girl, but I noticed that although her hair was done up in a bun on the top of her head as though for bathing, she did not seem to be wet.

'Why have you no clothes?'

'My father he cut them off me.'

By now I was fully awake, and offered to escort her to the police, but this she firmly refused, and asked to be put ashore. *Tofua* had berthed by then, and brightly lighted was lying at the wharf. I said that if I landed her there the girl could, if she so wished, speak with the policeman at the gate, or go to friends. She agreed, so we got into the dinghy, but I had rowed only a few yards when she suddenly cried out:

'Oh, no, you must not land me there, I have no clothes.' I stopped rowing. 'Take me to that other yacht,' pointing to *Roulette*, 'my girl-friend is there.'

I didn't believe her, for I knew Lawson pretty well by now, and I felt sure he would not welcome this sort of present at that time of night. So I rowed to the east, and selecting a dark place on the shore, turned, and backing the dinghy in with rather a creepy feeling, landed our visitor on the beach, quite expecting to be ambushed by an irate boy-friend or murderous father.

Looking back on that night I see there were several other and better things we might have done, such as offering her some clothes and a cup of coffee, or a berth for the night, or I might have accepted her final suggestion. As we neared the shore she leant towards me, smelling nicely of frangipani, and putting a hand on my knee said in the best Mae West style:

'You come with me?'

Well, what would you have done?

We sailed from Apia in mid-August, and having passed between the large islands of Upolu and Savaii, where little ferry-boats, which looked more suitable for use on the Upper Thames than on an open stretch of the ocean, were scuttling to and fro (a few weeks later one was lost and all her passengers drowned), we headed for Suva in the Fiji Islands. For most of this comparatively short hop of 640 miles we had fine weather and often were able to carry a cloud of canvas. On the way we passed close to the inhabited volcanic island of Niua Fo'ou, which has no safe anchorage, but did provide a convenient check on my navigation.

Fiji comprises 250 islands besides many islets and reefs. The most direct route from Samoa to Suva is through the Koro Sea, which lies inside the Lau (or Eastern) Group. There are several passages into the Koro Sea, but most of them are difficult because of the reefs which bound them, and the one most commonly used is Nanuku Passage at the north-east end; there, on

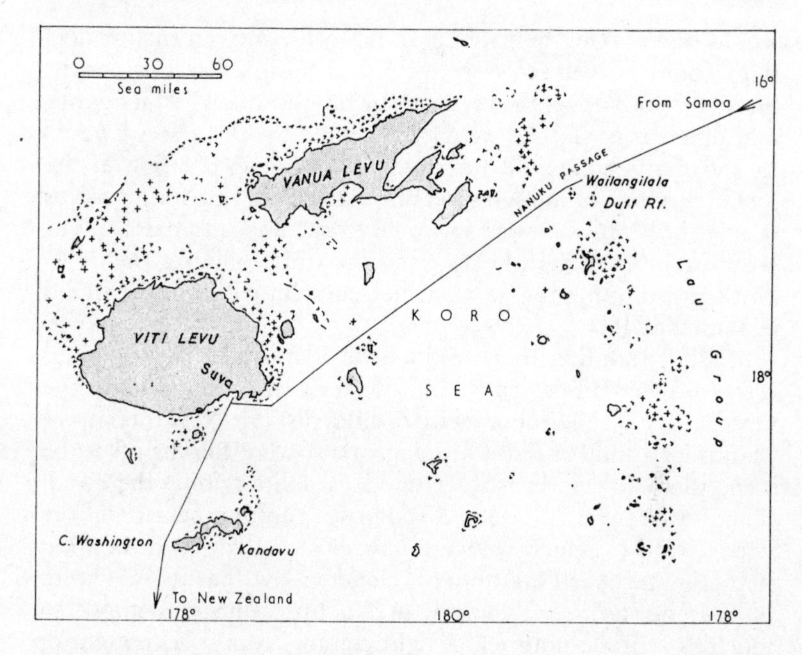


the low coral island of Wailangilala, stands a 15-mile light, the only one in the whole of the 160-mile-long Lau Group. But unfortunately Duff Reef, the only visible part of which is a tiny sand cay 4 feet high and with no vegetation on it, lies 14 miles to seaward of the light, and makes the approach from the eastward dangerous by night—if only the light could have been put on that instead it would have saved many a navigator much anxiety.

On our fourth night out from Apia we were hurrying along under much reduced sail with a strong beam wind, altering course a little now and then to dodge the fast-moving squalls, each of which came with a jet-black cloud and slanting shaft of rain. It really was a wonderful night for sailing as we swooped down into the troughs with our bow-wave boiling up high on either side, and then rose buoyantly up and over the crests in a shower of warm spray, the power-filled curves of our sails silhouetted by the moon, which frequently rushed out from behind the clouds to illuminate the splendid scene. But we could not feel carefree, for somewhere not far ahead the swell would be breaking heavily on Duff Reef, which in such conditions could be neither seen nor heard until we were upon it. Shortly before dawn the reef was, by my reckoning, 10 miles off, and that was as close as we dared to go in the dark, so we hove-to and made coffee. At first light I got an observation of the moon, and one of the sun soon after. The resulting fix placed us 8 miles north of the reef; so we let draw, and an anxious hour later Susan's wonderful eyes spied the white pillar of the lighthouse just topping the palms which surrounded it on the island of Wailangilala.

Having rounded that bright little island and its attendant reef, we sailed all day in a south-westerly direction towards the next light, a 9-miler on a reef 105 miles distant. Now there were islands all round us, and although most were too far off to be seen, those to windward sheltered us a little from the swell, though there was still quite a rough sea running because of the strong wind to which we were now closehauled. At dusk a big island to starboard lay under a cloud so low that its hills were buried; mysterious it looked, and a little sinister without so much as a single pinprick of light coming from it. I thought of the people there in their thatched, mat-floored *bures*, the beat of the pestle pounding the yanggona root, the clapping of hands round the kava bowl, the plaintive cry of a child, the whine of mosquitos; I almost wished we could have stopped there, but in Fiji it is important to go first to a port of entry because the officials wish to make sure that one's vessel is not carrying the beetle which destroys the palms from which most of the islanders make their living. So on we sped until before first light we sighted the outline of a high island fine on the port bow and much closer than it should have been, though there was still no sign of the light for which we had been heading. So

▶

23. Hauled out at Auckland. *A*: The original arrangement at the stern: a semi-balanced rudder hung abaft a huge space in which a 3-blade propeller turned; the spoilers, fitted at Grenada, can be seen. *B*: The balance part of the rudder has now been removed, and Peter is welding on one of the two steel plates to fill in part of the propeller aperture. *C*: To remove the filler with which the bottom had originally been thickly plastered we had it sand-blasted. *D*: The final arrangement. Two-thirds of the aperture have been filled in, a 'rudder post' has been fitted, a trim-tab has been added to the trailing edge and a plate to the upper edge of the rudder; the 3-blade propeller has been changed for a smaller one with 2 blades.

A ▲ C ▼ B ▲ D ▼

again we hove-to to await daylight, which showed that during the night either our compass had an unsuspected deviation, my allowance for leeway had been too much, or a current (or tidal stream perhaps) had set us to windward of the course—poor old navigator; if only he knew which, he could allow for it on some future occasion. With a compass bearing and a vertical sextant angle of the island's highest hill we fixed our position and then sailed on.

Yesterday had been Sunday. But during the night we had crossed the 180th meridian, so it was necessary to skip a day; today therefore became Tuesday, and (as on two earlier voyages) we were to have that rare thing, a 6-day week. This may seem confusing to those who have not experienced it, so perhaps a word of explanation might not be out of place. As a ship moves to the westward, noon grows later by one hour for each 15° of longitude she sails. When she is half way round the world from Greenwich, i.e. on the 180th meridian, noon with her is 12 hours later than it is at Greenwich, and if she completed the circumnavigation and did nothing except continue to put her clocks back one hour in each 15° time zone, she would return to Greenwich 24 hours late on Greenwich time. So when crossing the 180th meridian (or the date line, which is a modification of it enabling all islands in one group to have the same day of the week) she must skip a day if westward-bound, or add a day if bound to the east.

That forenoon we passed through a cluster of islands, giving their invisible reefs, some of which reached out 5 miles in our direction, a cautious berth, and then in clear water headed for a light off the low, reefy, south-east corner of Viti Levu, on the south coast of which lies the port of Suva. We rounded the light at dusk and held on towards Suva. But the night was windy and heavily overcast, and the town lights were so dazzling that although on earlier visits we had gone in by night, this time we preferred to wait for daylight. But while hove-to we made more

◀

24. *Top*: Rafted up with cray-fishing boats in Second Cove, Fiordland, during a spell of wild weather. Caddy McEwan, *Towai*'s skipper, scrubs her deck. *Bottom*: The best anchorage we found in Fiordland was near the Lake Alice waterfall where, as usual in that area, all but the sheerest bits of land were covered with dense, almost inpenetrable forest.

offing than we intended, and it was well after sunrise when we threaded the pass through the reef into the big harbour, and the business of the day was in full swing by the time we neared the wharf. We had been told that the boarding officers would no longer come out to the anchorage, and we were wondering where to go when the pilot cutter hurried towards us, and an immaculate, white-uniformed figure on her foredeck hailed us.

'Welcome to Suva, *Wanderer IV*.'

This proved to be the harbourmaster, Captain Harrison, who, like Bill Sellers at Lahaina, had recognized our ship from a picture he had seen, for no doubt she did look a little different from most other yachts in those waters. I asked where we should go. 'Right in front of my office,' he replied, 'but I'll go in first, then you come alongside me.'

Meanwhile a launch from the health department had been hovering near by, and as soon as we had tied up she came alongside. Her people must have thought from our welcome by the harbourmaster that we were some sort of V.I.P.s; before they asked us to fill in their forms they produced a bowl of kava (yanggona), and standing in a circle in our big cockpit we all drank in the ceremonial Fijian manner with the clapping of hands—fortunately Susan and I remembered that no matter how full the coconut husk is, or how full one may happen to feel, the proper thing is to drink its contents to the dregs in one go.

John Harrison, a sailing man who was having a Maurice Griffiths *Lone Gull II* built in a local yard, and his wife Bette, were good to us during our stay; they entertained us ashore and arranged for us to have lunch aboard and be shown round the big cable ship *Cable Enterprise*, which was based on Suva and covered the Pacific. I was interested to learn from her navigator that when, perhaps to locate the end of a broken cable, they needed to know exactly where they were, they rarely used the modern radio aids with which the ship was provided, but took a round of star sights—I wondered what they did in the North Pacific.

For most of our stay we lay off the Royal Suva Yacht Club, where we could get hot showers and cheap meals. Instead of walking into town to do our shopping, as we had in the past, we went by outboard dinghy and found we could leave her safely within a few yards of the market and not far from most of the shops. As well as an ocean-going ship or two we usually found lying at the wharf a bunch of small, characterful, inter-island trading vessels smelling richly of copra; some were so deeply laden that it seemed remarkable they could make their often rough trips with any degree of safety. As we passed, the people aboard these and any others working on the wharf always gave a wave and a flashing smile, and at the landing someone usually came running to give us a hand to get ashore. Never once did we see the outstretched palm which is the usual sign of the tourist trade, for although Fiji had plenty of visitors, at that time their advent had done little to interfere with the customs or spoil the natural good manners of the inhabitants as it had in some other islands. We noticed, too, that the Indians, who form so large a proportion of the population, looked better fed, better dressed—some of the women's saris were quite lovely in their grace and colour—and far more cheerful than on our last visit ten years before. When Susan went to the sawmill to get wood-planings for Nicholson's heads she was given a sackful, put in a truck and driven back to the club with a broad smile, and when she went looking for a sod with young grass, for Nicholson liked to graze when at sea, a small boy took her by the hand and said 'You're nice'.

During our stay the club gave a remarkable party for its members, their friends, and the visiting yachts. No tickets were sold, no money was taken, but everyone could (and many did) have as much to eat and drink as they could possibly stow away. Our rather lonely track from the north had at last joined up with the more frequented one across the South Pacific from Panama, and among other yachts lying at Suva it was good to find Ronnie Andrews's *Merlin*, the first British yacht we had seen in seventeen months. She, like ourselves, was bound towards New Zealand.

We left that busy, cheerful, colourful place in the third week of September, and soon sailed out from under the cloud which, as it so often does, was that day covering the eastern part of Viti Levu, into brilliant sunshine and a sparkling sea. During one of Susan's watches we rounded the western end of Kandavu in the night, and she watched the moon rise straight out of the top of the truncated cone of remarkable Mount Washington. There we altered course for Cape Brett in New Zealand, 1,000 miles to the south.

Optimistically we had hoped to carry the trade wind for a few days more, but unfortunately lost it soon after the land had dropped out of sight, and thereafter had winds from all over the place, varying from the lightest of airs to a good force 9, and for much of the time we could not lay the course. Often the sea, which for so many months of our voyage had been a deep, almost purple blue, turned grey under the leaden sky, and the cabin thermometer tumbled down from its accustomed 80-odd degrees F. through the 70's into the lower 60's, so that winter clothes, smelling strongly of mildew, were needed, particularly for the night watches, for since leaving San Francisco we had passed right through the tropics and were coming into higher latitudes again; but at least the nights grew shorter as we made southing. Even when the wind was steady in direction it was curiously unsteady in strength, sometimes changing from force 2 to force 5 and back again in a few minutes, and on those occasions we rarely had the right amount of sail set for either force, and therefore did not make good progress. In Fiji they had told us that we ought not to make this trip during the equinox as there is always a spell of bad weather then. I had found this difficult to accept, and said that were one to take heed of all such items of local weather lore one would not get very far. But on this occasion at least they were right, and although it did not bother us much there certainly was very bad weather farther south. At that time we saw in the sky, when it partly cleared, the portents of a depression: cirrus radiating from the west, and a solar halo, while at Wellington winds of 110 miles per hour were reported. We were to get our bad weather a little later, though nothing like so bad as that. It was two days after the equinox that the wind freshened so much and the sea became so rough, that after trying various combinations of sail in an abortive attempt to get *Wanderer* to steer herself, we finally took in everything, switched on our brightest lights, and both turned in for a rest. No sooner had we done so than we experienced a tremendous squall which, even though we had no sail set, hove us over to a surprisingly steep angle, and fairly screamed in the rigging. Heavy rain came with it, and continued through the night. Our radio receiver is said by its makers to be 'tropicalized'; however, its tuning mechanism is driven by string belts, and so damp was the atmosphere that these belts shrank and we could not tune in to get a weather report without risk of damage to the set. But the lightning which accompanied the storm would have made reception difficult, so we unplugged the aerial and went to sleep, and slept surprisingly well considering the conditions.

After breakfast the rain stopped and the wind took off a bit, so we set some sail and got moving again, but only for a short time, for when the wind backed to head us badly off our course we again stopped for more suitable weather. The radio, which by then we had dried out with silica gel, told us that Marlborough in New Zealand had been declared a disaster area due to the hurricane-force winds and floods caused by the heavy rain. Perhaps next time—if there is a next time—we *will* wait in Fiji for the equinox to pass before heading south.

It was not long before we stirred ourselves and got moving once more, but four days passed before we could ease sheets and sail at a reasonable speed in the right direction. By then another depression was moving east across the Tasman Sea, and we soon began to feel its effects; the wind shifted to the north, making a dead run for us, and as it strengthened the sea quickly got up, and it became even more difficult than usual to hold a steady course. At first I could not understand why this should be, for although we had not run dead before it in so rough a sea or rolled so wildly for some considerable time, we certainly had done so in the past. But presently, when I happened to steady the ship's head for a few moments on a tiny patch of blue sky over the bow, I noticed that each time we rolled the compass card swung about 10° first one side then the other of the course, and until then I had naturally been trying to follow the compass, thus steering a zig-zag course. Heeling error, which is caused by the athwartships movement of ferrous metal beneath a compass as a vessel heels or rolls, is one of the problems arising with steel construction, and this was our present trouble. Twice the compass had been corrected for this error by professional adjusters who fitted a vertical magnet immediately under the centre of the compass bowl. But the error varies with the latitude, and unless the seaman possesses a heeling-error instrument he cannot eradicate it himself. One of the professionals told me that when/if we crossed the magnetic equator—which we did shortly after leaving Fanning Island—the vertical magnet should be turned the other way up, and I tried that but without any noticeable improvement. Incidentally, a similar magnetic field affected the auto-helmsman's heading unit, and caused the equipment to turn the wheel to and fro as the ship rolled although she might still be on the correct course.

After some hours of wild steering, and even though taking only short tricks at the helm, we grew very tired; once again—and you might well think this was becoming a habit of ours—we hove-to for a rest under the close-reefed mainsail. Continuous rain had reduced visibility, but as we had seen only two ships since leaving Fiji, we felt the risk of collision was small, and we both slept well.

We awoke refreshed in the early hours to find that the sky was filled with stars and that the wind had moderated. When I went forward to unroll the reef I was astonished to see that several feet of the lower part of the sail track had been torn from the mast and broken, leaving a badly mangled end into which it would have been impossible to insert slides without first trimming it with a hacksaw. However, as we always used a lacing instead of slides for the lower third of the luff, no slides did have to be inserted, and I soon had the sail fully set. The trouble was caused because the forward end of the boom was rather far from the mast, so that when the sail was close-reefed the lower of the slides put a considerable pull on the screws holding the track to the mast. The lower part of the track should, of course, be mounted on a tapered wood chock to lead it aft from the mast to a point where its lower end is over the end of the boom, and in New Zealand a friend kindly did this for us so that we never had that trouble again.

We made full sail, and on a broad reach with many birds around us hurried along for where I hoped, for I had not been able to obtain sights for the past two days, Cape Brett lay. But, alas, an early observation of the sun showed that we were 20 miles farther offshore than I had supposed, distance lost, perhaps, while fore-reaching too much when hove-to. Our noon position was 25 miles east of the cape, so if we continued at our present speed we would be up with it by 1600, and at anchor in the Bay of Islands before nightfall. But the hours passed, and still there was no sign of the cape, which is the highest and most conspicuous piece of land in the neighbourhood. Not until 1700 did we sight it faintly through haze, and it was then at least 15 miles away.

Fifteen miles is a big error to make when one has had good observations, and looking back on my inefficiency I can only suppose that I must have misread the chronometer by one minute; this was possible as the minute and second hands of that ancient timepiece were not quite synchronized. But whatever the reason, it was by then too late to reach an anchorage in daylight, and as the haze was thickening, we spent the night jilling about off the cape, and came to an anchorage in the sheltered bay on the south side of Motuarohia in time for breakfast.

Since leaving Yarmouth in the Isle of Wight we had made good 17,500 miles along our crooked route to New Zealand.

# 6

# Improvements to the Steering

For two peaceful weeks we remained embraced in the soft, green arms of the lovely Bay of Islands, enjoying the serenity and dealing with an accumulation of mail. Mostly we lay at Opua, where Dick and Pat McIlvride had lent us the mooring just off their little scarlet-roofed house on Tapu Point. The proper approach to the house was from the water, and the front door faced the beach; the other, and less used approach was through a steep, rich meadow where sheep grazed. The McIlvrides were away sailing to Honolulu, and the Kennedys, a nice family from Ireland, were living in the house. John Kennedy was building a yacht, which was to be their future home, in the adjoining boat-shed, and kindly did some little jobs for me; Ann, his wife, gave Susan the run of her washing machine and invited us to use the shower; and each morning we watched with pleasure the sea-manlike, curly-red-headed children set off by dinghy for the school across the estuary. We made much use of our dinghy too, for over at the root of the wharf to which the occasional ocean-going ship came to load butter and mutton, stood a useful little store and post-office. Some of the people who read a story about us in the *New Zealand Herald* remembered our earlier visits 10 and 17 years before; we received many letters and telegrams of good wishes, and some of our visitors drove more than 100 miles to see us.

The fortnight passed all too quickly and we were reluctant to leave; but we had a major refit to attend to, and Auckland, about 130 miles away, was likely to be the most suitable place for that. However, as New Zealand was to be our headquarters for some time to come, we promised ourselves we would return presently for a longer visit, and sailed away; several days later, having stopped at some of the well-remembered harbours along the coast, we went in past the extinct volcano Rangitoto to Auckland Harbour and came to Westhaven, the chief yachting centre. It was crowded, but the custodian, Phlap Martinengo, and his wife, greeted us cheerfully and found us a berth close to the harbour bridge. Phlap—it is said that if he likes you, you can stay as long as you wish, but that if he does not like you, you leave Westhaven within 20 minutes—was busy fitting out his own yacht, and for that purpose had placed her right alongside his house so that he could step straight on to her deck from the upstairs veranda. In the near-by haul-out area there was great activity among the yachts with owners and crews completing fitting out, for New Zealand, it seemed, was still very much a 'do-it-yourself' country; apart from a few dinghies we saw no two yachts exactly alike, and this was refreshing after the large fleets of plastic one-designs we had seen in other countries.

Sometimes the New Zealander is a little rough and ready—'She'll be right, mate'—and he does have a rather engaging habit of 'improving' things; but some of the beautiful and efficient designs of high-ranking yacht architects, have, in amateur boatbuilders' backyards, been 'improved' beyond recognition. One man I met at Westhaven owned a really dreadful little cutter, with lean bows, fat quarters, and a broken sheerline, and insisted that I must surely recognize her. I tried, but failed, so he told me proudly:

'She's a replica of your *Wanderer II*.' The only similarity I could detect was that both yachts had bowsprits and transom sterns.

While one could not help but admire the New Zealander's independent determination to do things in his own way, it did seem wasteful that he was so disinclined to learn from the experiences of others—perhaps because he did not read much—and this was particularly noticeable in big issues such as pollution and traffic congestion. At a time when other countries were realizing and trying to correct some of the mistakes they had made, New Zealand was going steadily on to make the same old mistakes all over again. Where other towns abroad were trying to reduce pollution of the air, Auckland was about to scrap her clean, swift and silent trolley-buses and replace them with diesel buses; where others had realized the need to bypass small towns before they were completely ruined by traffic, New Zealand was widening roads and building bridges to bring more and more traffic through them. The trouble probably was that with a population of only 3 million as against Japan's 110 million on about the same area of land, there was so much space, air and beauty for everyone to enjoy, that it was not realized how valuable those assets were or how quickly they could be irretrievably lost.

Even in the minor yachting scene one saw a similar disregard of others' experiences. In Europe and the U.S.A., for example, most people were well aware that a yacht cannot get light and air into her accommodation through a skylight without getting water in too. Maurice Griffiths knew this many years ago and devised the double-coaming hatch to prevent the otherwise inevitable leaks, while others made exclusive use of opening ports in the coachroof coamings. But the New Zealand yachtie went happily on building massive old-fashioned wooden skylights which in rain or spray had to be covered so that they provided neither light nor air.

We needed skilful and understanding steel-workers to make the alterations we wanted done to *Wanderer*'s rudder and underwater parts aft in the hope of improving her abominable steering qualities, and were recommended to go to the yard of Vos & Brijs, which had built many fine wooden yachts in the past but now was concentrating more on steel fishing-vessels. When I learnt that the active partner, Len Brijs, came from Holland I had some qualms, for hitherto I had not got along too well with some of his countrymen. But I took to him the moment we met, and he proved to be a knowledgeable, efficient and kindly man. The provisional estimate he gave for the work we wanted done—this included sand-blasting and repainting the bottom—was very reasonable; but he warned us that he could not be sure of the cost until he saw the ship out of the water, and that the final bill might be considerably higher. I asked if we would be permitted to live aboard while on the slip; he replied that he hoped we would do so as the yard was open and unguarded and there could be a risk of theft.

We were hauled out in November and the steel work was at once put in hand. As I have mentioned before, *Wanderer* had been built with a semi-balanced rudder abaft a large gap in which a 3-blade propeller turned (Plate 23*A*), and I had reached the conclusion that although this might be an excellent arrangement for a fully-powered craft it was wrong for a sailing vessel. Our plan was to fill in much of the gap (one could hardly call it a propeller aperture as it measured 5 feet by 2½ feet), to cut off the balance part of the rudder which projected 5 inches forward of the stock, and to fit a snug, streamline rudder post, though perhaps that is not the correct term to use as it was not intended in any way to support the rudder, but merely to smooth out the flow of water on its way to the rudder. As we might one day want to withdraw the propeller shaft, the fill-in would have to be done with two plates so that a gap could be left between their after ends just forward of the rudder, and this gap was to be bridged by two pieces of timber bolted to one another through the fill-in plates, thus making an easily removable 'rudder post'. Our American friend Henry Chatfield had suggested all this and had made a drawing showing how it could best be done. An additional refinement was to have a small trim-tab welded to the lower part of the rudder's trailing edge; this was to be made of steel sufficiently thin to be bent while the ship was afloat as trials might show to be necessary, until it eventually eradicated the hull's inherent desire to turn to port, something which until now the spoilers had been attending to, but of course we wanted to scrap them because of the drag they set up. It was also our intention to fill in the gap between the top of the rudder and the underside of the counter to prevent the escape of water there when well heeled and sailing fast. This plate was to be bolted to the rudder so that we could remove it easily if it proved to be of no value or, by increasing the area of the rudder, should make turning the wheel too hard. Finally, we had decided no longer to drag while sailing the big, 3-blade propeller, but to change this for a 2-blade one of greater pitch; then we would be able to lock the shaft with the blades vertically abaft the sternpost, in which position they would cause the minimum drag when sailing.

The slip at the yard of Vos & Brijs sloped at an angle of 5°, and as *Wanderer*'s keel is parallel to her waterline, she remained at that angle, which may not seem much, but we found during our first night there that we could not sleep until we had jacked up the feet of our bunks with piles of books and blankets; while we were doing this we had a surprise visit from a smart policeman from a prowl car, who wanted to know just what we were up to on board. During the day the unaccustomed angle brought some little-used muscles into play, and they ached. Also we discovered when we washed down the deck that all of the water could not escape overboard at the stern, but stood on the afterdeck and leaked through the flush-fitting hatch into the after peak. Masking tape could not put up with this, so we bailed out the water and prayed for fine weather.

While the upper of the two steel fill-in plates was being welded to the hull there could have been a risk of fire on board; we therefore ripped out some of the joinery-work in the cabin so that I could stand a fire watch there. I felt a bit nervous as I watched the inside of the plating glow bright red, saw the paint blister and burst into flame; choking in thick, blue smoke I got down on hands and knees, scraped the burning paint to one side and beat out the flames, repeating the operation many times. But there was, I thought, an even greater risk of fire that day. The crew of an A-class keeler lying close astern and on the same slip as ourselves, had decided to paint her bottom with one of those paints which smell strongly of acetate; on the can was a notice in red to the effect that it should not be opened within 60 feet of a naked flame. The paint was being sprayed on, and to protect Peter, our welder, from the fine mist of paint, our stern had been enveloped in an old, very light spinnaker, which billowed in the breeze; I suspected this of being inflamable. As Peter, with his hissing, spark-throwing torch worked on unconcerned, I suppose I need not have worried, but I was indeed thankful when the welding was finished and I could rebuild the cabin woodwork.

While the alterations to stern and rudder were being made, some of our many visitors expressed doubts about their likely success, and bluntly asked if I really knew what I was doing. After they had gone away I used to go back on board and have another look at some extracts I had taken from Skene's *Elements of Yacht Design*. One of these read:

'Sailboats should have unbalanced rudders, that is, the axis of rotation must be at the forward end and attached to the keel. It has been found that balanced rudders hung independently, well aft of the keel, steer poorly and increase resistance. The uniform flow of water, so necessary in way of rudders, is broken up, but, by the introduction of a narrow skeg immediately ahead of an unbalanced rudder excellent control is obtained'.

Other experts, including Douglas Phillips Birt, had made similar comments, and I had learnt that by placing a snugly-fitting fin (my 'rudder post') immediately forward of the rudder, the angle of stall might be increased from 12° to 30°. The 'rudder post' was made of two pieces of tallow-wood, which it is said no worm will attack; incidentally, it does not float.

Some visitors thought we were crazy to change to a 2-blade propeller, and told us this would produce unbearable vibration and lack efficiency, but I consoled myself with the thought that if this proved to be correct we could always change back to the old 3-blade propeller. There were conflicting comments on the type and brand of paint we proposed to use, each owner of a steel yacht recommending something different; some people said we had not enough sacrificial zincs, others that we had too many. It was all very interesting but somewhat confusing.

Meanwhile the yard was in a state of chaos, for a new building to house offices and canteen was being erected across the very limited space ahead of us. One day a mobile crane came to lift the steel framing on to its foundations, and then a portable diesel engine to drive the tools sprang into noisy life. At one time *Wanderer* had to be let back down the slipway a few feet as she was in danger of having her bowsprit built in to the second storey. The yard was so littered with debris of all sorts that I found it difficult to get around without injury. If I looked where I was putting my feet I was likely to split my head open on some higher hazard, and if I looked out for that I might break a leg. Then, to add to the general uproar, when our steel work had been completed, Arthur King's sand-blasting crew arrived with their own noise-making machine, another diesel engine on a trailer. The blasting was done wet, i.e. a high-pressure jet of water from a big-diameter hose into which a supply of sharp sand was continuously fed was played on the hull (Plate 23*C*); this made a special noise of its own, a penetrating roar. Stan and Norm—we were all on a Christian-name footing there—who took turns with the hose, wore oilskin suits and face-masks. The reason for all this was that against our express wishes the builders had coated the entire hull with some form of plaster or filler to conceal any imperfections; much of this had already fallen off the bottom leaving areas of bare, rusting steel, and the sand-blasting was intended to remove the rest of it together with rust and any paint, and burnish the steel before it was repainted. We ought to have had the topsides done at the same time, but could not afford the expense. Some of the filler, which looked and felt something like linoleum, defied the sandblast and had to be peeled off with knives, thus lengthening the operation and presumably adding to the cost. But by knocking-off time the job had been done, and the mess was almost indescribable: on deck and below we were covered with sharp grit and fine, clinging, brown dust, much of which had come up through the secret scuppers which I had forgotten to plug. The yard and most things in it were in a similar plight. During the night a thin film of rust formed on the bare plating, so starting early next day, Stan and Norm did the whole bottom over again in preparation for the first coat of paint, and this time the sand was used dry, so the cloud of dust was even more penetrating and we were to find traces of it for many days to come. But that job did not take long, and at the end of it the plating gleamed almost like silver. The Intertar, a two-pot paint, was being prepared when someone noticed that at a weld below the turn of the bilge liquid was trickling out in three places. This proved by smell to be diesel oil from our fuel tank which was about half full, and was leaking from faults in the weld where, presumably, only paint and filler had held it in before. Peter was called from some other job to re-weld the seam, but was unable to do so because the leaking fuel at once ignited and burnt his hands, and I was hopping round like a fussy old hen with chicks, saying that the tank would explode or the ship catch fire. Attempts to weld had to stop; the painting gang stopped mixing paint, and a council of war was held; the upshot of this was that the proper thing to do was to empty the tank of its remaining fuel, and try again; meanwhile—and it would be a long while, for the only way to empty that tank, which probably had 100 gallons in it, was by using the small sludge pump with which it was provided—no other work could well proceed. But Peter—he also came from Holland, and like Len Brijs had lived in New Zealand for the past 15 years—thought of a better plan, and proceeded to carry it out. Quickly he drilled a hole at each of the three leaks; these he tapped and fitted with threaded plugs; he ground down the plugs and then was able to make a proper job of the faulty seam. Having attended to that he took Susan's favourite 10-year-old-saucepan away and replaced its broken handle with a new one, and then made a long pinch-bar for me so that I could 'dive down' and alter the angle of the trim-tab as necessary—the pinch-bar must have weighed at least 12 lbs, and fortunately I had the sense not to go swimming with it, as it or both of us would have sunk like a stone.

That day Susan and I joined forces with Stan and Norm to put on the first of two coats of Intertar, and found our workmates interesting and entertaining people. We had noticed while hauled out at Kettenberg's yard at San Diego that the yard workers, for whom we paid the yard $10 an hour, did seem to get on with the job and not waste much of their time or our money talking. In New Zealand we paid about half that sum for labour, and everyone appeared to have plenty of time to talk and laugh and have tea breaks, and generally enjoy their carefree and independent way of life; yet we gained the impression that in the end just as much work got equally well done in the same space of time. Susan continued to paint with our mates next day, but I did not as I was having a session with the N.Z. agent for our steering gear, who had come 140 miles from Tauranga to replace a ball-race and check the installation.

After the final sand-blasting we could not wash down during the rest of our time on the slip because it was important not to wet the steel or any of the intermediate coats of paint, so the ship remained in a shocking state of filth, and so did we, for although the office staff had made us honorary lavatorial members, shower facilities were not yet available. We trusted that the people who might have seen us on television in such an unkempt state understood the reason, and we hoped we did not sound too gritty in the broadcast we were asked to make while living on the slip. That our personal standards had fallen low will be appreciated when I mention that one evening Susan decided to have a 'good wash' in a bucket; later, while doing her hair, she discovered that she still had a carpenter's pencil tucked behind her ear.

Living aboard while on the slip is in some respects like being in hospital. Friends may bring gifts of fruit and magazines together with their good wishes, but only the experts can really help. We received plenty of gifts, fruit certainly, and bottles of wine, and Arthur King—perhaps feeling that his blast of sharp sand had not been the best of introductions—brought us soup and scones, and presented us with a huge pork pie and a magnificent square Christmas cake, which his wife had made, and on the top of which he had made a sugar icing picture of *Wanderer* sailing into a colourful sunset.

We helped put on the first of the two coats of antifouling paint which followed the Intertar, while Arthur put in an appearance now and then to make sure with his magnetic paint gauge that we were getting it on thick enough; the four of us took only 1½ hours. But the final coat Susan and I put on unaided, the day being a Sunday, and we were anxious to save expense. We took 6 hours! In the process of sand-blasting and undercoating we had lost the waterline, and although we had in advance taken measurements down from the rail capping in many places, the prospect of having to cut in a new, fair waterline with masking-tape was daunting, particularly as most of it was out of reach, and the only planks and trestles available were too heavy for us to lift. The evening we were looking at this job and wondering how best to tackle it, our friend Peter Cornes, part owner of the big motor yacht *Sirdar*, arrived in clean, white trousers and suede shoes offering to take us for a drive in the country. This we declined, but as he is tall and understanding about yachts, we persuaded him to run the tape while we stood as far back as we could to correct by eye any highs or lows. On the starboard side there was just room enough for this, but as a shiny, corrugated-iron building, on which the evening sun beat hotly round *Wanderer*'s dark shadow, stood within a few feet of the port side, we could not help much there. However, Peter made a good job of it, and my log has a note to the effect that with this and other paint lines which had to be cut in cleanly, we used a total of 600 feet of masking-tape during this refit.

After eleven mercifully rain-free days on the slip, during which the new building ahead of us grew and grew, we were launched and went back to our berth at the Westhaven marina, where we had a grand scrub down with fresh water, and cleansing hot showers, and there we remained for three weeks enjoying much hospitality and completing our refit. We have always found New Zealanders very helpful, and among others to help us here was John Brooke, Commodore of the Royal New Zealand Yacht Squadron, who fitted the auto-helmsman's motor with the new armature its makers had sent out from England to replace the badly worn one, and made our 'irreparable' (according to the instrument makers) barometer as good as new; also he got his son-in-law, who runs a boatyard, to make our afterpeak hatch watertight by cutting a recess in its under side where it touched the steel coamings, and filling this with a strip of soft rubber; it never leaked again. We spent much time shopping in the city, and as Susan had made a new weathercloth for the cockpit and wanted to fit it with brass eyelets, tried all sorts of shops and some of the sailmakers for the necessary tool, but without success until one day we discovered Fosters, the ship chandlers. To a young assistant who was wearing an apron over his shorts I said:

'It may seem a silly thing to ask, but do you happen to have a dolly in the place?'

'Yes,' he replied without hesitation, 'what size do you want?'

Fosters did not stand in the best street, neither did it seem to go in much for chrome-plated fittings or electronic equipment, though I suppose one could get such things there if needed; but from its dim, tar-scented Aladdin's caves, one of which I recall contained anchors of many types in a large range of weights, it rarely failed to provide our nautical wants.

Meanwhile we had been waiting with some apprehension for the yard's bill, for a lot more work had been done than was covered by the original estimate. But when it came we found that both the yard and Arthur King had been generous—the final bill came to just two-thirds of the estimated cost.

Having got Captain Keane, a compass adjuster, to attend to our steering compass, which he found had acquired 20° of deviation (due presumably to changes in the ship's magnetism) since it was last adjusted in California, we sailed away on trials out in the Hauraki Gulf, and found a wonderful improvement in *Wanderer*'s behaviour. Before the alterations had been made she was so unsteady on the helm that one had frequently to apply half a turn, or more, of the wheel to keep her on course; but now she was much steadier and responded quickly to a spoke or less. She showed a slight tendency to turn to starboard instead of to port, which proved that the trim-tab was working but needed to have its angle slightly reduced. We found no undue vibration from the 2-blade propeller, and no noticeable loss of efficiency when going ahead, though the astern gear did not act quite so efficiently as a brake as it had done when swinging the 3-blade propeller; but these trials were carried out in fairly quiet weather, and we had yet to prove the worth of the alterations in the open sea. Unfortunately, however, the rudder was daily growing stiffer and harder to move, probably because the wooden 'rudder post' was swelling and binding; so we returned to Westhaven and went on one of the scrubbing berths. There were several of these in that harbour, each consisting of a stout, one-plank-wide jetty against which to lean as the tide dropped, and with blocks alongside to take the keel; each had fresh water laid on, for ease of scrubbing.

Although the tides were springs we did not dry out, and it was a deep paddling job to unbolt the two pieces of tallow-wood so that Len could take them back to the yard to be eased. By the time he returned with them the tide had risen a lot, and Susan and I were up to our chests in water as we bolted them on. We also took this opportunity to use the pinch-bar and reduce the angle of the trim-tab a little. We floated off that evening. went to a mooring for the night, and next day set out on a trip which we hoped would take us down to and right round the South Island, something we had dreamed of doing and had looked forward to for many years.

At the south-west corner of the South Island lies a mountainous area which, except for the peaks, some of them snow-clad, and cliffs too steep for vegetation to cling to, is covered with dense, almost impenetrable forest. Its coastline faces the stormy Tasman Sea where hard, onshore winds prevail, and into the deep, steep-sided, glacier-gouged chasms of this remote and virtually uninhabited region the sea thrusts a dozen crooked fingers of dark, cold water with charted depths in excess of 100 fathoms; though some of these penetrate more than 20 miles inland, anchorages are few, and only the heads of two of the inlets could be reached by road. This area is known as Fiordland, by all accounts a wild, exciting, lonely place, though an historic one, and Susan and I were determined to visit it during our proposed cruise round the South Island when we left Auckland ten days before Christmas.

The previous year Dick McIlvride and John Guzzwell had made a cruise in company round the Dominion, and Dick had lent us many of the charts we would need. When I bought new charts to fill in the gaps and make the set complete before using and returning it to him, I noticed that most of them were no longer produced by the British but by the New Zealand hydrographic office. These were in the modern style with the land and some of the sea coloured, and although mostly based on British Admiralty originals, they sometimes lacked the detail which was given with such loving care and accuracy on the latter. We noticed that some names had been changed (this is in keeping with British Admiralty practice) from English to Maori; for example, Step Island, which is remarkable because of its series of steps, had become Motukawaiti, Cliff Island had been changed to Whangarara, and it looked as though the Ninepin was about to become Tikitiki, though in this instance the old name was still given in brackets. The point arises: are these charts intended for use by English-speaking or Maori seamen? If for the former, which seems most likely, then English words, especially when descriptive, should be used as much as possible, for anything that can make the navigator's job quicker or more certain should be provided, and it does not help him when trying to identify some feature to have to refer to the glossary given in the *Pilot*; also it is far easier for him to pronounce and remember an English rather than a foreign word. The latest supplement to the

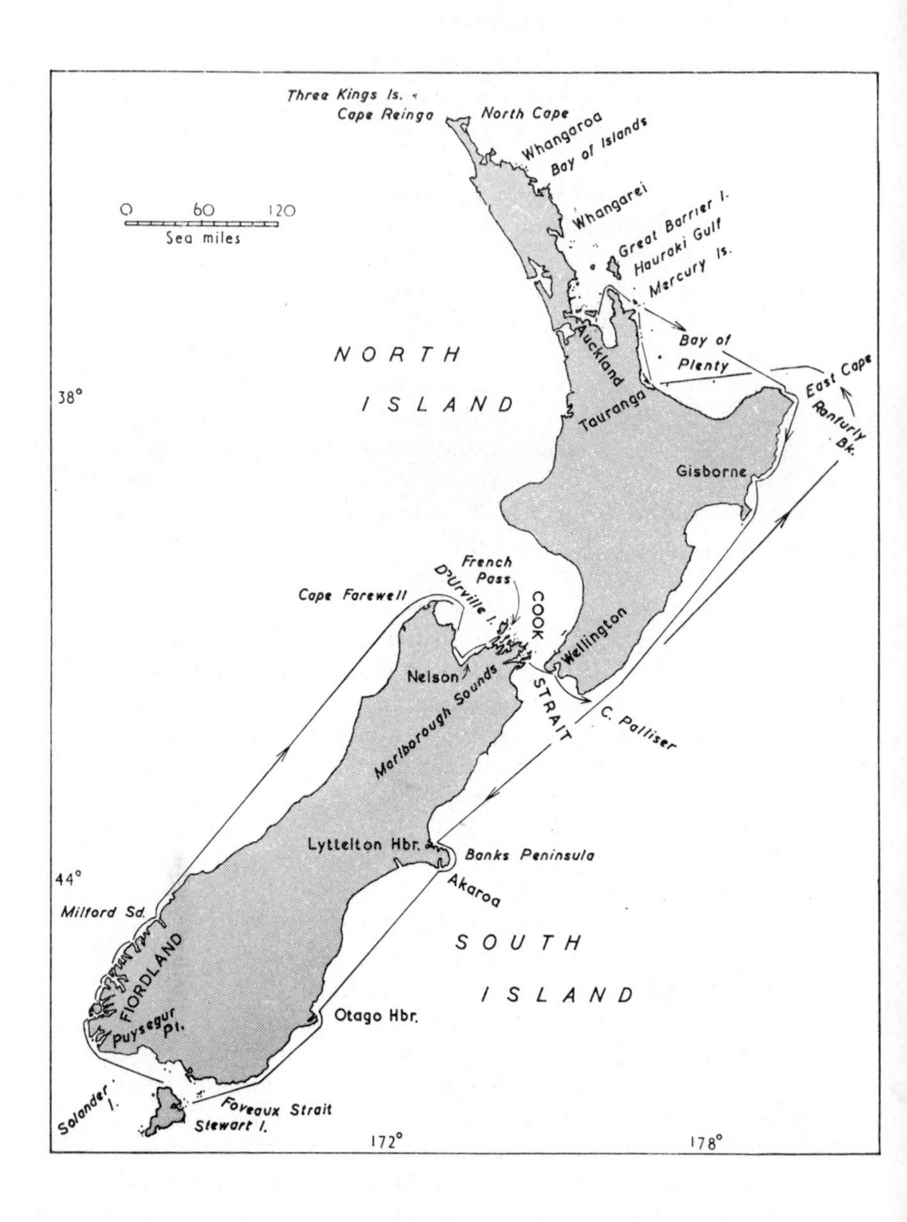

Three Kings Is.
Cape Reinga
North Cape
Whangaroa
Bay of Islands
Whangarei
Great Barrier I.
Hauraki Gulf
Mercury Is.
0 60 120
Sea miles
NORTH
ISLAND
Auckland
Bay of
Plenty
East Cape
Ranfurly Bk.
Tauranga
Gisborne
38°
French Pass
D'Urville I.
Cape Farewell
COOK
STRAIT
Wellington
Nelson
Marlborough Sounds
C. Palliser
Lyttelton Hbr.
Banks Peninsula
Akaroa
44°
Milford Sd.
FIORDLAND
SOUTH
ISLAND
Puysegur Pt.
Otago Hbr.
Solander I.
Foveaux Strait
Stewart I.
172°
178°

*New Zealand Pilot* had not caught up with the change of names or with all of the new chart numbers; this we found unhelpful and confusing.

To reach Fiordland we had a choice of three routes: we could go up round North Cape and down the west sides of both the North and the South Island; but the west coast of the North Island, with the great surf created by the recurring eastward march of depressions across the Tasman Sea, has little to attract a yacht, for most of its harbours have dangerous bars. We could go down the east side of the North Island, then through Cook Strait and down the west side of the South Island, but if the weather was true to pattern we could have a long, hard beat from Cook Strait on. So we chose the third alternative: to go down the east side of both the North and the South Island, where in the event of headwinds or bad weather there would be more and better anchorages in which to seek shelter. This route has the serious disadvantage that it is usually difficult to round the southern end of New Zealand west-about against the prevailing Roaring Forties headwinds and foul current; but at least Stewart Island, which we wished to visit, would make a convenient stop in that area. If we succeeded in reaching Fiordland we planned to return to Auckland by way of the west side of the South Island, where we ought to have a fair wind and plenty of it, then through Cook Strait, which would give us the opportunity to visit the Marlborough Sounds and Wellington, and then back up the east side of the North Island. The total distance would be in excess of 2,500 miles, and this is perhaps one reason why so few yachts manage to reach far south; their owners do not have the time, or if they have they prefer to visit the warmer and less stormy islands of Fiji or Tonga.

The weather was quiet and fine as we sailed gently away from Auckland to find an anchorage for the night in one of the many sheltered coves on Waiheke Island's southern shore. After the continuous roar of traffic over the harbour bridge, close to which we had been lying, the silence was delightful. Fortunate indeed is the Auckland yachtsman in having this lovely place and the sheltered waters of the Hauraki Gulf with its gentle tides and wide choice of anchorages right on the doorstep of his city.

We stole away early in the morning, and leaving the gulf

astern, headed out past Cape Colville to another peaceful anchorage at Great Mercury Island. It is rare to find a rock-girt island in open water having a sheltered harbour with good holding ground; Great Mercury, which was feeding a flock of 5,000 sheep, is a notable exception; its harbour is almost a creek. After one night there we pressed on to round East Cape while the weather was fine and the wind fair, though there was very little of the latter when we started on the 130-mile crossing of the Bay of Plenty; however, it breezed up early in the night, and the sea quickly grew rough. Owing to a defect in its heading unit, the Pinta was inoperative on this cruise except when steering courses with large north or south components, and for the greater part of it we had to steer by hand. But since the alterations we had made to the rudder were so satisfactory, this was no longer the physical struggle it had been in the past; we felt that Vos & Brijs's magic welding torches had changed *Wanderer* almost overnight from a virago to a gentle lady, and all through that lovely, moonlit night she hurried on her way light and easy on the helm. But in the morning, as we approached East Cape, the wind shifted quickly to the south, and as we did not wish to beat against it down to Gisborne, the nearest harbour, we turned into Hicks Bay, which is 1½ miles square, has depths of from 5 to 7 fathoms, fine sand, and appeared to offer excellent shelter.

Soon after we had anchored *Wai-iti*, a small crayfish boat, came over from a jetty at which she had been lying to advise us that if the south wind freshened we should move close in under the cliffs fronting the south side of the bay where it would not be felt. We ought to have taken that advice and moved before dark, but as the wind was not strong we were content with our berth in the middle of the bay.

At about midnight a violent squall from the south woke me, and going on deck, where it was raining hard, I was dismayed to find that we were lying broadside to the wind although there was no tidal stream—a clear indication that we were dragging. I therefore called Susan, we weighed, and very slowly under full power moved over towards the southern shore. As the bay carries its depth right up to the rock ledges which front the cliffs, sounding was of no help, and in the dark, the sky being heavily clouded over, we failed to get in close enough to find shelter from the gale, which continued to worry us with tremendous squalls. We took it in turns to keep an anchor watch, and in each squall it was obvious that we were dragging, and by 0130 had lost so much ground that again we had to weigh and motor to windward. Twice more we repeated this performance. At 0300 *Wai-iti* found conditions too rough to remain alongside her jetty. So she cast off, came past us and anchored close under the cliffs, thoughtfully leaving her lights on as a guide to us. We then weighed for the fifth time and were able to get in close to her where, just as her skipper had said, we were quite sheltered from the wind. Later we learnt that Hicks Bay is notorious for its bad holding ground, and that shortly before our visit a large trawler had blown ashore there while her crew were asleep. However, as we dragged in other places later, we came to the conclusion that our 60-lb CQR (plough type) anchor either was not heavy enough or was defective, and decided to replace it with one of 75 lb as soon as possible.

By mid afternoon the wind had moderated and shifted a little to the west, so having eaten the three crayfish which *Wai-iti* had given us, we left and managed to round East Cape, where under the grey sky the sea was green and angry with the tide, just before dark, and made an uneventful passage down the coast. Here, as in so many parts of New Zealand, there are some off-lying dangers, but these were not marked, except perhaps by breakers, for buoys are not used except in the approaches to commercial ports.

The following afternoon we came to Gisborne, where the leading beacons take one almost on to the beach before the harbour opens up; we went into the basin where we were given a hand to tie up alongside a fishing vessel which was temporarily out of commission, so we were able to lie there undisturbed by the comings and goings of the early morning fishermen.

When Cook made his first New Zealand landing near Gisborne there was nothing there except a river with a shallow, shifting bar, and as he was treated badly by the natives and failed to obtain any of his needs, he called it Poverty Bay, a name which today does not suit this fertile and productive area. Recently a wall was built in the river, dividing it into two longitudinally. Then, leaving the river to do what it wished in its western part, the engineers dredged the eastern part, where there was now no tidal stream to cause silting, and constructed the snug basin. This was rather crowded towards the end of our stay, for although the weather was fine with a clear sky and little wind, so heavy a swell was running on the coast (this was said to be due to an earthquake in Ecuador) that the fishing fleet could not tend its pots, and at near-by Napier the plates and frames of the British freighter *Canopic* were damaged before she could be got away from the wharf.

# 7

# New Zealand—a Cruise round the South Island

Christmas day found us crossing the southern approach to Cook Strait. Through this the winds funnel often with great strength, and as though to prove the point the beam wind, which had been light all the way from Gisborne, suddenly increased to 40 knots in the early hours of the morning. In the open sea and well away from the land we would probably have stopped in such conditions to wait for an improvement, but here we felt it might well go on blowing hard for some time, and the best thing was to get across the mouth of the funnel as quickly as we could. So we took the jib and mizzen in, reefed the mainsail, and under that and the staysail alone we tore along; the gale whistled in the rigging, the spray rattled like hail on the weathercloths, and every now and again when the bow-wave collided with a breaking crest, a big burst of it drove across the cockpit. However, Susan at the wheel was well protected by the helmsman's dodger, and from the shelter of the companionway, where I was struggling out of my oilskins, I could just see the top of her souwester appear for a moment above the dodger as she lifted her head to scan the horizon for the lights of other ships, and then disappear as she again looked down at the compass. It was a vile night to leave her there alone, but I knew she was confident and competent, and was probably enjoying herself. I tried to sleep, but the noise and the motion and the urgent feeling induced by our speed prevented this.

I took over at 0400 and watched as the blackness melted slowly, and when at last dawn came it revealed a dismal scene: grey sky above a grey-green, windwhipped sea streaked and patched with white. Cape Palliser, the nearest land probably 25 miles away, was invisible. Our appetites for breakfast were not great, but our roaring progress was heartening, and every hour the spinning log clocked up another 6½ sea miles. There were no ships to be seen that Christmas morning, but there were plenty of birds, including the magnificent albatross, though this was not the wandering but the mollymawk species, soaring and banking with rarely a movement of its great wings; and there were some fine black-back gulls.

In the forenoon the wind began to moderate and, so sudden are the weather changes in those waters, by evening we were almost becalmed under full sail again and slamming about in the left-over sea with barely steerage way. We enjoyed our supper, which was mostly of the things the kind people of Gisborne had given us—Susan counted twelve different types of fresh food on board—and we washed it down with a bottle of California wine. No doubt this made us idle and drowsy, and soon after we had finished it we agreed we had had enough of trimming the sails to take advantage of the vagrant airs with which we made hardly any progress; so we took in all sail except the mainsail, which we sheeted hard amidships, lashed the helm, switched on the stern-light, hung the riding light on the forestay, and both turned in.

At dawn, on looking out to starboard (although without steerage way, *Wanderer* optimistically still pointed in the direction we wished to go) we saw the jagged line of the two-mile-high Kaikoura Mountains, their snow-capped peaks showing pink and insubstantial in the early sunlight against the pale blue of the sky. An hour later we were beating against a freshening headwind, wet and despondent, for Lyttelton, our intended destination, lay nearly 100 miles away at noon. But the wind did not last, and in the evening, being by then becalmed again, we started the engine and kept it running all through the windless night so that we picked up the loom of the lights of Christchurch in the early hours, and in the morning entered the harbour. Although this looks impressive on a map, it is not the perfect haven one might suppose, and only in the walled, commercial port is good shelter to be had from all winds. But except with northerlies the bays on the south side offer fair shelter, and we anchored in one of these called Diamond Bay, where we were snug and content and almost overwhelmed with hospitality.

I have never understood why nearly everyone we met in New Zealand was so kind and generous to us, for surely a yacht from overseas is not uncommon. At times we felt sad that we could do so little in return. On our last day in Diamond Bay we took the ferry across to Lyttelton and went by train to Christchurch to do some shopping. On returning in the evening we wished to prepare for sea so as to be able to make an early start in the morning. As we disembarked from the ferry we found waiting at the landing place a Christchurch family which had taken the day off to come over and see us; they had brought gifts including strawberries and home-made jam. Obviously they were longing to come aboard and talk, for they were about to build a large, steel ketch for long-distance cruising. But it was nearing the end of our busy day, and we felt we had to get on with the business of making ready or else stay another 24 hours; so to our shame and regret we did not invite them on board. Perhaps we are in too much of a hurry, always striving towards some distant goal; perhaps we are too selfishly busy and occupied with our own affairs; I know that I am too impatient, and perhaps one day I will relax and take life more easily, have time to sit around and yarn; perhaps . . .

We did get away as planned at dawn and sailed round Banks Peninsula in company with several of the small, silver Hector's porpoises, to the harbour of Akaroa, where we saw many red patches, each several square yards in area, on or near the surface of the water. From a distance these had the appearance of growths of kelp marking rocks; at closer range they might have been blood from some large, wounded creature; but in fact they consisted of millions of tiny frog-like creatures which, we understood, were whalefeed.

Continuing on our way south we allowed one mile in ten for the expected indraught towards Ninety Mile Beach. Unlike others of the same name, this beach really is of that length, and because of the low, flat ground behind it—the patchwork quilt of the Canterbury Plains—can be a danger to navigation at night. We had an uneventful, 160-mile trip to Otago which, like Lyttelton, looks impressive on a map, but the chart shows that although it is wide and 12 miles long, most of it is occupied by shallow or drying mudbanks, and offers no really snug berth for a visiting yacht. We felt we needed a secure anchorage just then because cyclone Rosie with 60-knot winds was reported to be on her way south, and we wished to remain in port until she had made up her mind what she was going to do. We sailed in through the well-marked entrance with a fair wind, passed the only mainland nesting ground of the albatross, and the bay where Robin Knox-Johnston in *Suhaili* ran aground during his single-handed voyage round the world, and having neared Port Chalmers, which is about half way up the harbour, we anchored just outside the fairway near the fishing vessels' berths. There was no point in going into one of the several vacant berths, as they consisted only of piles and a walkway, and offered no protection from the short sea raised by the wind blowing up the harbour. However, the holding ground appeared to be good, and that evening we learnt with relief that Rosie was filling up.

We waited a day for the very low glass to rise a bit, and the time passed quickly as a nice young television crew from Dunedin came with sound-on-film equipment to 'do' us. They came at 1100, and we all had such an hilarious time (plenty of wine with lunch) that they stayed until teatime.

As we motored away on the last of the ebb in the early morning calm we wondered what the day would bring, for we had now reached the point where we must start to round the bottom end of New Zealand, a coast which offers no sheltered anchorage in 130 miles, and it was there that we could expect headwinds, rough seas, and a difficult or, as some had found before us, an impossible trip from which we might have to return defeated, and that had happened to fully-powered seagoing fishing vessels.

So we could scarcely believe our good fortune when, having come out of the harbour and round Cape Saunders, where we felt the lift of the southerly swell, we picked up a fair wind. To this we set a running sail to help balance the mainsail, and with a cloudless sky overhead, whitecaps all around, and many birds busy fishing, we watched the rounded underbelly of New Zealand with its trees all leaning to the east to show which way the prevailing wind blew, slip swiftly by. It was one of the most lovely sailing days imaginable; it was even warm; but our progress was almost too good, for the eastern approach to Stewart Island, which lies south of the southern extreme of the South

Island, and is separated from it by Foveaux Strait, is guarded by rocks and islands with which we had no wish to close during the hours of darkness, for although the southern end of the South Island is well enough lighted, the approach to Stewart Island is not lighted at all. So in the late afternoon we took in the running sail and the mizzen to reduce our speed a little, but this was not effective for long because the wind started to freshen.

At the conclusion of her watch Susan called me at midnight to say that we had run our agreed distance and now should be about 10 miles south-east of the outermost danger towards which the tidal stream would be setting for the next four hours. So we hove-to heading south-east and probably drifting south. While keeping an intermittent lookout during my watch, I wrote up my journal on the chart table under the shaded red light while the blue flame of the galley stove hissed as it heated water for the coffee. I noted that Bob Griffiths and his party in the ferro-cement cutter *Awahnee* on an attempt to sail round the Southern Ocean in 100 days along the 60th parallel, and Chay Blyth in *British Steel* going the other way, were presumably somewhere to the south of us. Apart from the fringes of Antarctica, where several countries have bases, the tip of South America, and a few scattered islands, there was no inhabited land in latitudes higher than our own, and that just then was 47°S. To people who live in the northern parts of Europe and America 47° may not seem to be a very high latitude, for after all the English Channel, Newfoundland, and Vancouver Island lie in 50°N., but the Southern Ocean has no Gulf Stream or North Pacific Drift to keep its temperature up, and no great land masses to temper its westerly winds or check the seas; therefore these can blow and run unimpeded on their way.

Most of the night was fine and brilliantly starlit, but before dawn the entire sky became overcast so quickly it seemed as though a blind had been drawn across it, and drizzle set in. At first light we let draw and ran on towards Stewart Island, a misty lee shore with each of its protecting reefs and islands—the Mutton Birds, the Fancy Group, Bench Island, the Twins—standing in a ring of foam. For a time we suffered the tense anxiety which is attendant on the approach to any strange place in rough weather, and is heightened when running down on a lee shore; but as usual the landmarks were eventually recognized and fell into place, and at about 0900 we passed the last of them and ran into the smooth water of Paterson Inlet, which almost cuts the island in two, and anchored off the boat harbour behind a small islet in a bay on the north shore. Had the wind been from the westerly quarter we would have brought up off Oban, the only village on the island, and the southernmost inhabited place in New Zealand; but that day the Oban anchorage was off a lee shore. However, the village was only half a mile away, so we went there to collect our mail from the strangely modern post-office—somehow we had expected this far southern village to consist of a huddle of stone cottages, but instead found a sprawl of widely scattered, brightly painted, iron-roofed bungalows. And as the island had only 3 miles of unsealed track, we had certainly not expected to see taxis tearing along in clouds of dust to and from the air-strip. We soon shifted to a secluded anchorage in Glory Cove at the south side of the inlet, but had several tries before we could get the anchor to hold there because of kelp. With steep, heavily wooded hills on all sides of us we found absolute silence and solitude, and on a small pebble beach near-by Susan discovered a spring at which to fill the watercans and have a washing day; so clean and sweet was the water that we topped up our tanks with it.

I have referred earlier to the steering error which can arise when a grid type compass is affected by dip. It appears to be the custom of some compass makers to put a small weight on the card to compensate for this; but such a weight, placed for use in the northern hemisphere, aggravates the problem in the southern hemisphere. At Stewart Island, where the angle of dip is 72°, our compass card was so badly tilted out of the horizontal plane that I glued lead weights to the verge ring so as to tilt the bowl and minimize the steering error. Later the makers sent us a new magnetic assembly counterbalanced for the southern hemisphere; this improved matters, but when next we cross the magnetic equator we will have to replace it with the original assembly. I know of no other disadvantage of steering by a grid, and it has many things to recommend it. Incidentally, it may be of interest to note that the magnetic variation near Stewart

Island, which was 22°E. when we were there, had increased by 6° in the 200 years that had passed since Cook first sailed those waters and recorded it as 16°E.

The island, which is mountainous and 670 square miles in area, has seen the comings and goings of shipbuilders, sealers, and whalers, while timber mills have risen and fallen into decay. As most of it is thickly covered with bush there is hardly any land fit for farming, and the 500 islanders, all of whom live at Oban, today make their living by fishing, and by catering for the few visitors who fly in or go over by ferry from Bluff.

We had left our island anchorage at first light to catch the tide in Foveaux Strait, where it runs hard, and just managed to get through that turbulent stretch of water before the stream turned against us. All through the afternoon solitary and uninhabited Solander Island, sentinel of the famous and dreaded but now deserted whaling ground, grew in stature on the port bow as we ran with full sail set to a fair wind, and by teatime we had it on the beam. We were indeed fortunate to have such favourable conditions in this notoriously stormy area, and as the glass was high and steady and the sky was clear, we trusted that the weather would remain good at least until we had got round Puysegur Point, the south-west corner of the South Island, for that headland has a bad reputation on account of its sudden weather changes and the violence and frequency of its gales; observations taken at the lighthouse over a period of 50 years show that gales have blown on 14 days in every 100 throughout each year.

We raised the light, a bright flash every 15 seconds, soon after nightfall, and with a failing wind crept on to round it, and twice did I have to call Susan from her watch below to help me gybe as the fitful breeze veered and backed. But apart from the lack of heart in the wind the night was wonderful, crisp and clear, with not a cloud in the sky; and the moon, just past the full, shone brilliantly and lit up the white cliffs of Chalky Island to give us a check on our progress some hours after we had put the light astern, as we headed along the coast for Dusky Bay.

Perhaps a word on nomenclature might not be out of place here. In most parts of New Zealand Maori names naturally predominate, but along the coast of the remote south-west most of the names are those which were given by the early European explorers. Although the area we were now approaching is known as Fiordland, the term 'fiord' is not used for any of the inlets; most are called 'sounds', and this is confusing because of the Marlborough Sounds, which are far away and have no connexion with this area; but two of the Fiordland inlets are called 'inlets', and one, Dusky, is called a 'bay'. In 1770 Cook in *Endeavour* sighted this part of New Zealand but did not stop, and I quote from his Journal for Wednesday, 14 March:

> 'In the P.M. had a fresh Gale from the Southward, attended with Squalls. At 2 it clear'd up over the land, which appeared high and Mountainous. At ½ past 3 double reeft the Topsails, and hauld in for a Bay, wherein there appear'd to be good Anchorage, and into which I had thought of going with the Ship; but after standing in an hour we found the distance too great to run before dark, and it blow's too hard to attempt it in the night, or even to keep to Windward; for these reasons we gave it up, and bore away along shore. This bay I have named Dusky Bay.'

We approached Dusky, or what we hoped was Dusky, as the sky was changing from dark blue velvet to pale blue silk, and the rampart of mountains that defends it from approach by land was sharply silhouetted by the sun which had not yet topped it. This, our first glimpse of Fiordland, was exciting, and both of us felt a tingle of slightly apprehensive anticipation, for this was to be our cruising ground for the next two weeks, and we had heard of some of the problems it poses for small craft. The chief of these is the difficulty of finding water shallow enough in which to anchor, for the fiords are very deep, often with no bottom at 100 fathoms, and with very deep water right up to the shores. Such few anchorages as there are usually drop off steeply into deep water, and for this reason it is prudent after anchoring to run a stern-line to the shore.

It was hard at first because of our distance off, haze low on the sea, and the lack of perspective to the back-lighted mountains, to decide where the entrance lay; but as we stood in cautiously towards the shore the landmarks resolved themselves:

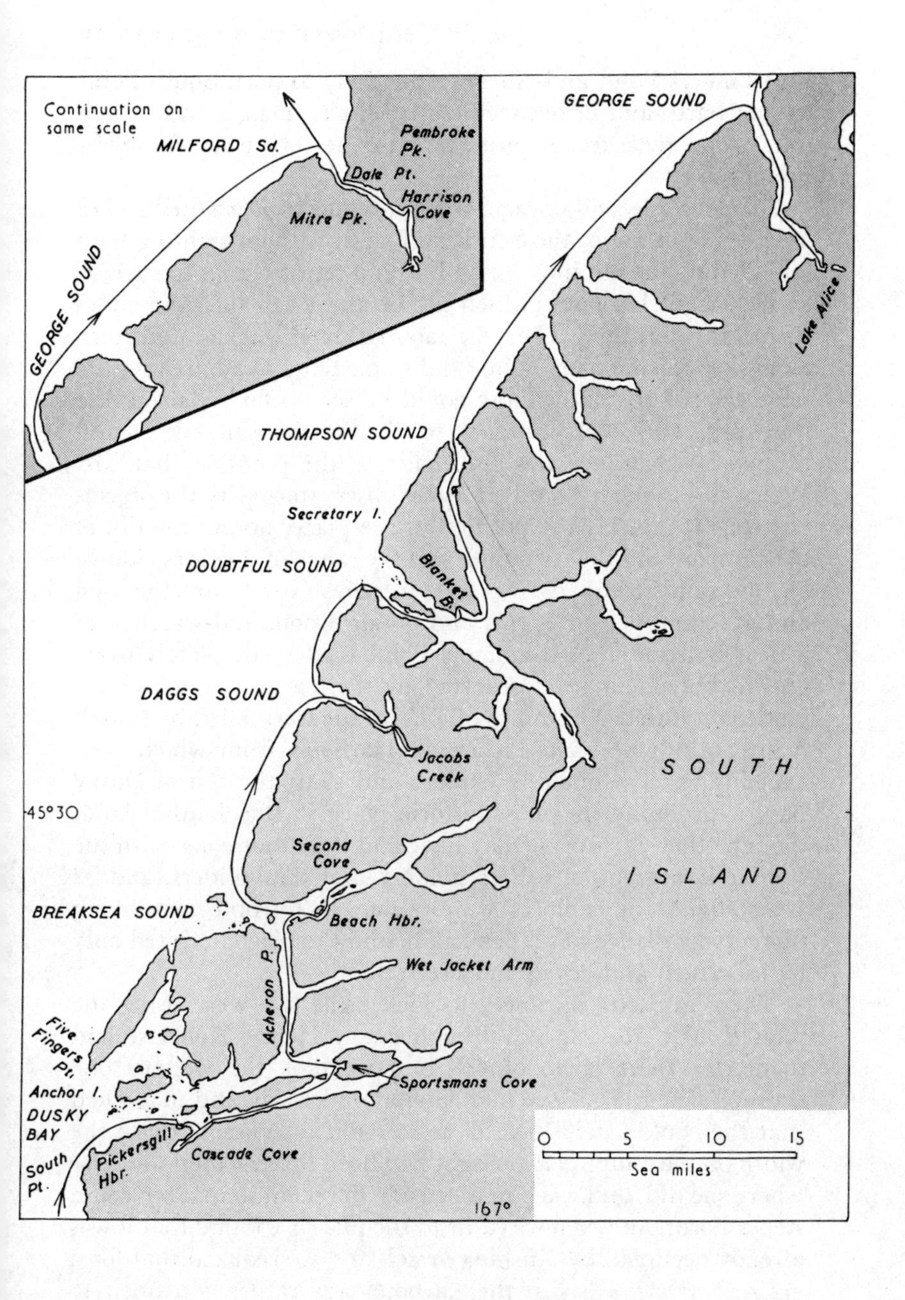
Continuation on same scale
MILFORD Sd.
Pembroke Pk.
Dale Pt.
Harrison Cove
Mitre Pk.
GEORGE SOUND
GEORGE SOUND
Lake Alice
THOMPSON SOUND
Secretary I.
DOUBTFUL SOUND
Blanket B.
DAGGS SOUND
Jacobs Creek
SOUTH
ISLAND
45°30
Second Cove
BREAKSEA SOUND
Beach Hbr.
Acheron P.
Wet Jacket Arm
Five Fingers Pt.
Anchor I.
Sportsmans Cove
DUSKY BAY
South Pt.
Pickersgill Hbr.
Cascade Cove
0 5 10 15
Sea miles
167°

Five Fingers Point and Anchor Island lay to port, South Point to starboard, and in between clustered the small islands which offer some measure of protection to the bay from the ever-present swell.

On Cook's second voyage, when he came in *Resolution* in 1773 after being at sea in the Southern Ocean without sighting land for 123 days, he put into Dusky Bay and remained for five weeks at Pickersgill Harbour, which lies on the south side four miles within the entrance. There he moored his ship head and stern alongside Astronomers Point, and immediately had areas of the bush cleared so that a forge could be set up for repairing the ironwork, and tents pitched to shelter the sailmakers and coopers. A clearance was also made on the point so that Mr. Wales, the scientist, could set up his observatory with the objects of fixing accurately the position of the place, noting the dip of the compass and the variation, and the range of the tides. Above all, the ship's company was able to refresh itself after the long and gruelling period at sea, and obtain wood, and water from a near-by stream. Astronomers Point is now completely overgrown, but it has been reported by the few people who have landed on it that the stumps of some of the trees felled by Cook's men can still be seen. Pickersgill Harbour, from which boat expeditions were made to explore and chart the rest of Dusky Bay, is probably the most historic place in the South Island, though other parts of the bay have had their moments, with the coming and going of sealers, miners, and shipbuilders, and 22 years after Cook explored it more than 200 people were living there. Now silence has fallen again, and the place is visited only by fishermen and occasional deerstalkers.

There are two entrances to Pickersgill (we were using the plan Cook's 15-year-old draughtsman, Henry Roberts, had made, for there is no other), and we took the western one through which *Resolution* had entered under sail, and marvelled that she could safely have done so for it is 'scarcely twice the width of' that ship. Our thought had been that we might anchor where she did, and warp in alongside the point so that we could step ashore; but when we came to the place we found that it was already occupied by a fishing vessel, and we realized that for a craft of *Wanderer*'s size the harbour was rather too open to easterly winds, there being a fetch of at least 5 miles in that direction.

By now the morning was well advanced, and we, growing a little tired, felt we would appreciate a snug anchorage. So we left Pickersgill by the wider eastern entrance, and turned south-west into Cascade Cove, at the head of which the chart showed we would find anchorage in from 5 to 10 fathoms. We passed the waterfall which gives the cove its name, and went on in towards the head. I remember it as being a sunny place and its vegetation a paler green than elsewhere; but this must have been just a trick of the light, for the trees—beech, rata and rimu—and the ferns and mosses, were similar to those seen everywhere else. We passed the beach where Drs. Charles and Neil Begg had pitched camp while they explored by dinghy to gather the material for their excellent book *Dusky Bay*; incidentally, that is one of the few places in Fiordland where there is a beach of any kind. But at the anchorage we found the soundings so irregular that we believed the bottom must have become foul with waterlogged tree-trunks and other debris, as has happened elsewhere in those waters; and as by then the wind had arisen and was blowing straight in, we decided that this was not a good place in which to stop. But could we find a better?

The chart showed that at the western end of Coopers Island, 10 miles farther up Dusky Bay, there was a landlocked basin called Sportsman's Cove. Against this on our borrowed chart someone had pencilled the word 'fuel', so we supposed it must be a place frequented by fishing vessels, and we set out for it.

The wind was freshening, and as soon as we had got out to the main arm of Dusky we found it was not blowing from north-east, as it had been in Cascade Cove, but was from the west (local wind variations such as this are typical of the fiords) and we ran inland at high speed. The shores of the bay and the islands within it are covered with dense forest, and when we reached Coopers Island we were for a little while hard put to it to discover the entrance to the cove, for it is narrow—much narrower than the narrow entrance to Pickersgill—and is kinked so that one cannot see through it. However, we did find it, and we ran in. With startling suddeness the bottom came up to meet us, with every rock and pebble and patch of weed clearly visible, and as there are no soundings on the chart, nor any information about depth in the *Pilot*, we did wonder if there would be water enough to let us through. This was a bit too exciting, but as the flood was just starting to make and there would be a rise of 7 or 8 feet, we knew that we were bound to float off no matter how hard we hit the bottom if we should run aground. However, the least depth we found in that astonishingly narrow and twisty channel was 4½ fathoms. With wind and tide hurrying us along we shot through it, and as it began to widen out into the basin, Susan and I looked to port and starboard expecting to see fishing vessels, or at least a tank or some oil-drums on the shore, appear from behind the trees. But there was nothing, nothing at all but steep, forest-clad slopes encircling the cove, which was about half a mile long and a quarter of a mile wide; below the trees stood a nearly vertical wall of black, shiny rock. The echo-sounder said 'No bottom at 50 fathoms', and not until we reached the far side and were very close to the shore where a stream came tumbling down, did we find a narrow shelf with 8 fathoms on it. We let go the anchor and instantly were attacked by swarms of sandflies.

This was our first serious encounter with those creatures, which are the curse of Fiordland. Of them Cook wrote:

> 'The most mischievous animals here [Pickersgill] are the small black sandflies, which are very numerous, and so troublesome, that they exceed everything of the kind I ever met with; wherever they bite they cause a swelling, and such an intolerable itching that it is not possible to refrain from scratching, which at last brings on ulcers like the small-pox.'

We had been bothered by *nonos* in the Marquesas, horse-flies in Australia, ants in Florida, wasps in the Great Barrier Island, and mosquitoes in countless places, but never had we seen the like of this invasion by Fiordland sandflies. We found that the insect-repellent we had with us was not effective for more than ten minutes or so; one could watch the creatures hovering over an anointed area of skin until the first effects wore off, and then come in to attack, and we wondered what they lived on when there were no humans around. We soon learnt that when on deck, particularly at twilight, we must wear long-sleeved shirts with button-up collars, and trousers tucked into thick socks, for the flies were partial to wrists, throats, and ankles. *Wanderer* is provided with portable screens for all hatches, skylights, doorways and opening ports, and after shipping these we had no more sandflies below except for the dozen or so which came in on our persons each time we entered. Mercifully the flies do not go far from the shore, and always they return to their homes for the hours of darkness. Before we left England our family doctor had given us a lot of medical stores including tubes of an ointment called Anthisan; application of this, we found, immediately stopped the 'intolerable itching' and enabled us to sleep properly without developing 'ulcers'.

During our years of voyaging we have spent nights at some strange, wild, and lonely places, such as Bramble Cay, Santa Fé, and Night Island, but somehow Sportsmans Cove (Cook's name for it as a boat expedition was made there from Pickersgill, and some birds were shot) seemed more lonely than most, and far more lonely than the open ocean hundreds, or thousands, of miles from land. When darkness came after a flaming scarlet and orange sunset, which was not so much beautiful as frightening, there was scarcely a sound; no creature stirred, though there must have been many round us; no fish disturbed the mirror surface of the inky water; there was absolutely nothing to be heard but the chilly tinkle of the little stream, the ticking of the French carriage clock in the saloon—its gentle chime was startling—and the beating of one's heart. Here, it seemed, was a place utterly separated from man and from any normal form of wildlife; silent and sinister. A wind in the trees might have brought it a touch of life; but perhaps it was just as well that there was no wind, for we, not yet familiar with the proper procedure in Fiordland, had omitted to run a line to the shore, and a strong puff of wind might quickly have blown us out into such deep water that the anchor would have had no hold. But in spite of our sense of foreboding the night was uneventful—no snakes climbed up the bobstay to paralyse us in our bunks, no giant octopus reached up to snatch Nicholson from his perch on the mizzen boom, no sea monster caused *Wanderer* to drag her anchor off the shelf—and I believe we both slept well. But after an early breakfast we were not sorry to motor away across that deep cove, and out through the crooked channel into the more open water of Dusky Bay.

We were bound for Breaksea Sound, the next big inlet to the north, but there was no need to go to sea to reach it, for Dusky and Breaksea are connected by Acheron Passage, a spectacular waterway 7 miles long and in places only half a mile wide, on each side of which mountains of from 3,000 to 4,000 feet rise from the water's edge so steeply that nothing much can grow on them. That morning the peaks were not wreathed in mist, as I think mountains should be to gain mystery and perspective, but were clearly etched against the ice-blue sky, and so absolute was the calm that they were perfectly mirrored in the glass-like surface of the water until our wake disturbed the reflection. Cook left Dusky by this route, which he called New Passage, and took several days to get through it; its present name was given to it 80 years later by Captain Stokes when he surveyed those waters in H.M.S. *Acheron*, for he had an unfortunate habit of changing the names Cook had given, and of altering Cook's charts, though some of his 'corrections' had to be changed back later to agree with Cook's findings. On we went past Wet Jacket Arm, off the mouth of which is a charted depth of 202 fathoms, and so into Breaksea, where we knew anchorage could be had on the south side at Beach Harbour, which is a channel inside some small islands.

After the emptiness of Dusky, Beach Harbour seemed almost a metropolis. On the tiny area of beach which gives the place its name, lay a dump of rusty fuel drums; a shack, presumably used by deerstalkers, stood near-by. There were two mooring buoys out in the middle, and at the shallowest place, where the harbour has a 6-fathom midway bar, a tiny, clinker-built fishing boat, *Flora*, lay deserted. Shortly after we had anchored the fishing vessel *DaVinci* came in, handed us a bucket of crayfish tails and then secured to one of the moorings. She was soon followed by two others, *Waikaremoana* and *Towai*, which spoke to us and then rafted up with her. Two of these fine, able-looking vessels were from the port of Bluff, the other from Stewart Island, and they made a gay splash of colour in that deep-green place. That evening we were visited by some of their people, who told us a little about their work. It seemed that most of the crays had been fished out of the sheltered inlets so that now the pots had to be set outside along the exposed coast, where almost always a swell is running, and where pilotage close in among the rocks must be done entirely by eye. Clearly this calls for steady nerves, high-class seamanship, a reliable engine, and skill in handling lines, for a fouled propeller could bring instant disaster. Most of the vessels had deep-freezers in which plenty of food was carried, for they remained on the coast for many weeks at a stretch, and into which the crayfish tails were put. Nearly all had two men aboard, but a few of the smaller craft were run single-handed. *Flora* was one of these, and only a week earlier had been found drifting near the rocks at the mouth of Breaksea with her engine still running, but with no-one on board. The other fishermen searched for her single-handed skipper for several days, but did not find him.

At dawn all the boats left and we followed them out, but soon met the early starters returning with the news that there was too much wind and sea. We continued on to have a look for ourselves (we had no pots to tend), but quickly agreed with them and turned back to Beach Harbour, where in the freshening wind we could not get our anchor to hold properly on the shingle bottom although we tried in three places. Peter Tait, skipper of *Waikaremoana*, seeing this, slipped from his mooring, went up harbour a little, dropped his big fisherman anchor with lots of chain and gave it a strong pull; he then bent a line and buoy to the chain and shouted:

'Here's a mooring for you, chum. Use it as long as you like.'

Thankfully we picked it up, but that evening all the fishing vessels got under way, and Caddy McEwan in *Towai* hailed us.

'Looks like she's going to scream from the nor'-west, and we might all get blown out of here tonight. Come on over to Second Cove.'

On the chart I had looked at that place, a bight on the north shore of the sound with a depth of 27 fathoms in it, and had not considered it further. But a little local knowledge is a valuable thing, so we slipped our mooring and followed *Towai*. Caddy, who had his son Gary as crew, went slowly so that we could keep up, and we were grateful to him, for by now visibility was much reduced by rain and it was growing dark. In the cove we found that the other two boats had already anchored and taken aboard the buoyed shore-line they had rigged up there for an occasion such as this, and were lying with their sterns close in under the trees. Caddy anchored and rafted up with them, and we then let go where he told us (on top of his anchor, we feared) and rafted up alongside him (Plate 24, *top*). That evening while we lay so cosy and secure with the nor'-wester roaring in the trees above our mastheads, we invited all hands aboard for drinks, and got to know them and a little more about their way of life.

All next day it rained and blew and we sympathized with our companions whose strings of pots were probably being smashed up by the sea outside. By evening three more boats had joined our raft, and again we invited all hands, twelve of them this time, for drinks aboard—a jolly party which lasted for five hours. Nobody showed much interest in listening to the weather forecasts, for no doubt they got their weather long before the meteorological office at Wellington heard about it; but they did listen, as did we, to the reports given every six hours from lighthouses—that evening Puysegur Point was recording 45 knots.

In those parts a north-west gale usually shifts to the south-west, and may then blow even harder, and our cove would be no good then. So the following evening as the wind started to shift, it was decided that we should all return to Beach Harbour. When we weighed we found, just as we had feared, that our anchor had fouled *Towai*'s. By the time we had cleared hers and dropped it again it was too close to the shore to be of any use to her, and we, with the wind blowing us broadside on into the cove, were in an awkward position. Peter, who was already under way, quickly sized up the situation and returned to tow us out stern first and then went back to give *Towai* a pluck. I felt my seamanship had not been of the best.

During the night the gale blew itself out, and at daybreak everyone put to sea to salvage what remained of their pots, and bait and lay new ones. *DaVinci* was the last to go, and her skipper came over to tell us that the sea had now gone down a lot. Before leaving he went to poor little *Flora* and pumped her out.

It was a windless and gloomy morning; for much of the time patches of drizzle hid the iron-bound coast from us as we motored north against a short, steep sea, which presaged wind to come from that direction soon, and rolling heavily in the beam swell. Twice, when visibility improved for a little while, we could see *Towai* working close inshore, apparently among the breakers. We had hoped to reach Doubtful Sound, 20 miles to the north, but had not covered half that distance when the expected wind sprang at us from dead ahead, and progress became so slow and wet that we turned to starboard, and with the wind then on the beam reached into near-by Daggs Sound. In the drizzle this looked dark and forbidding, but it is shorter than most, and at its head only 8 miles from the sea we came to what appeared to be a well-sheltered anchorage in Jacobs Creek surrounded by mountains which half hid the sky. The glimpses we had of *Towai* near the coast that day brought home to us the dangers of the fisherman's life, and we were relieved to see her come in some hours later, anchor further down the creek and pick up a shore-line.

That night it rained in typical Fiordland fashion (a fall of 30 inches a month is not unusual), and towards dawn we were disturbed by a roaring which we thought must be wind in the trees; but when day reluctantly broke we discovered that the noise came from waterfalls created by the rain, dozens of them, their silver threads streaking the rugged sides of the mountains. So hemmed in were we, and so dark was the day, that we needed the saloon lights on until mid morning. Caddy weighed and came over to tell us that the wind outside was blowing at 35 knots, and that if it freshened any more it could scream where we lay. He invited us to raft up with him and his neighbours, scarlet *Isobel* and rusty little *Star of the Sea*. All had haunches of venison lashed in the rigging; one gave us a big cut of this for our 'tea'; another gave us boysenberries and ice-cream from her freezer.

The following morning was foggy, but at first light the fishermen went about their business: *Towai* to work the coast back to Breaksea where, all being well, she would spend the night; *Star* to work north to Doubtful, and *Isobel* to fish the seaward side of Secretary Island. We felt a little lonely when they had gone, the more so as the fog thickened to keep us there among the invisible waterfalls for three more hours. Even then it did not clear, but lifted just enough to let us see the shoreline underneath. We crept out to sea under power, and found it clear there but with no wind—the all-or-nothing pattern of weather to which we were growing accustomed—so wc motored north to Doubtful where, we had been told, Blanket Bay on the south-east side of Secretary Island offered the best berth. There is a tiny islet in that bay which had a refrigerated depot where cray-fish tails could be left to be collected by a small seaplane summoned by radio-telephone; the fishermen found it more profitable to use that expensive form of transport than to make the long, usually stormy, trip round Puysegur Point to dispose of their catch at Bluff. We had just anchored and picked up the shore-line, when down came the dainty little plane to make a graceful landing and taxi to the island to load her cargo. A cheerful young man from the islet came off by speedboat to ask if we had any letters to post, which we had, and a few minutes later the plane was in the air again, heading inland up the deep chasm of the sound for Te Anau, the nearest town some 50 miles away; we watched the silver speck diminish and fade out of sight.

That evening scruffy little *Star of the Sea* came in and asked if she might raft up with us. Of course we agreed, for not only did she have prior right to the shore-line, but we knew and liked her crew. So she anchored and came gently alongside. Most of the fishing vessels we had met were well maintained and had comfortable accommodation; but poor *Star* appeared to be in need of a refit, and seemed very primitive; while her engine ran, water sluiced her rusty deck so that her crew had to wear seaboots, she was sadly lacking paint, and her only accommodation appeared to be in the deckhouse, a small, steel box, in which were two hard chairs and a coal-fired cooker. We were just going to sleep to the lullaby of her radio when the wind sprang up from the south-east, blowing straight into the bay with a fetch of 6 miles; as *Star* was shorter than *Wanderer* the two vessels did not pitch in harmony, and because both had

▶

25. As we approached Dale Point at the narrowest part of Milford Sound, there was a sudden lifting and lightening of the gloom ahead, and there before us stood the great mountains.

rubbing strakes it was difficult to keep fenders in position. By midnight the sea was rough and there was risk of serious damage being done. *Star*'s skipper, realizing this, said he would leave, but before doing so insisted on passing us the bitter end of his anchor cable because he knew we had dragged our own anchor in Beach Harbour. In exchange we handed him our spare 100-lb fisherman anchor and a nylon warp. Casting off from us he motored away into the black and windy void of Doubtful Sound. Now with no neighbour and with two anchors out to windward we were quite safe, but we felt uneasy about *Star*. I do not know where she spent the remainder of that night, and our anchor and warp had not been used when she returned them to us in the morning. Two months after this incident we were sorry to learn that *Star* had foundered one stormy night in Daggs Sound; her crew took to the dinghy, and our good friends, Caddy and Garry in *Towai*, managed to find and rescue them—a splendid achievement in the dark.

We were glad to get out of Blanket Bay and find a more peaceful cove off Thompson Sound, which is an alternative exit from Doubtful. After that we made only one more stop before reaching Milford; this was at the head of George Sound where we found a delightful anchorage, probably one of the best in all Fiordland, beside the cascade which takes the surplus water from Lake Alice, 200 feet up (Plate 24, *bottom*). There, perhaps because the sky was clear of cloud, the scene was less austere than in our earlier anchorages; the mountains, though high, did not press in on us so closely, the depth was only 5 fathoms, and the greatest fetch in any direction was only half a mile. We brought off a dinghy-load of water from the cascade, together with a swarm of sand-flies, and that was the first time we had set foot ashore since leaving Stewart Island.

Milford is the most spectacular of all the sounds, so it was a disappointment when we came to it next day after a 35-mile trip with a light, fair wind, to find the cloud down to a few hundred feet. As we came into the mouth of the sound the wind

◀

26. Night comes early among the mountains. While *Wanderer* still sparkles in the afternoon sunshine the western side of Harrison cove is already black and featureless.

freshened from north-west, and we ran fast before it. But as we approached Dale Point at the narrowest part, there was a lifting and a lightening of the gloom ahead; within a few minutes the cloud dissolved, and there before us stood the great mountains (Plate 25). It was like watching from a darkened auditorium a stage backdrop being illuminated. To starboard reared the stupendous cliffs of Mitre Peak; to port loomed a mountain almost bare of vegetation and which, from its peculiar colour, appeared as though made of metal. Beyond this Stirling Fall, fed by the perpetual snow on Pembroke Peak, came streaming down. But we had little time to enjoy all this magnificence, for we were running fast up a blind alley and wanted to stop in Harrison Cove before we got blown past it. Sails had to be stowed while we tried to identify the landmarks as they swept by. The beacon at the cove did not appear until we had passed it; we turned quickly to port, and within a couple of ship's lengths came suddenly out of the blast of wind and the white-caps into a calm pocket, and carried our way into dead-smooth water, and a little breathless anchored in 10 fathoms. We at once launched the dinghy, and while I paid it out Susan ran a line to the shore, and I had to bend together nearly all we had to make enough for her to reach a tree—100 fathoms in all.

This was a most spectacular berth with fantastic mountains on all sides of us, while high over our stern glistened the now sunlit snowfields of Pembroke Peak; to the east of us Bowen Falls sparkled against its dark background. As always when among the mountains night came early, but it came much sooner to the western than to the eastern side of our cove, for the latter was still colourful in sunshine while the other was black and featureless (Plate 26).

We had now reached civilization. A motor road ran to the head of Milford, where a river had been diverted and an airstrip built in its place. Near it a big hotel stood beside Freshwater Basin, and from its pier, much hampered by the pack of fishing boats that lay there, large motor launches took the visitors out to see the wonders of the sound. Some people think this is a pity, but Susan and I do not agree with them. We feel that it is proper that anyone who wishes to do so should be able to visit a part of Fiordland without having to own a yacht or a fishing vessel. Nevertheless we are glad that the other sounds to the south remain aloof and inaccessible, as silent, lonely, and unspoilt as they have always been, and that we had the good fortune and the satisfaction of seeing them from the deck of our own ship.

The day we planned to leave Milford Sound did not look very promising; cloud cover was down to about 200 feet, rain was falling, and although there was little wind where we lay, whitecaps were hurrying in the sound. However, by now we had learnt that the weather at sea is often quite different from that in the fiords, and there was the possibility that if we waited the wind might freshen more and make it difficult to get out. So after an early breakfast we cast off our sternline from the tree, got the dinghy on board, weighed anchor, and left our cove. In the main part of the sound we found the weather even less inviting. Only the feet of the mountains could be seen and, narrow though the sound is, at times no land at all was visible. Above the noise of the wind and the breaking crests could be heard the roaring of the waterfalls, over-fed with rain, as we steered a succession of compass courses, identifying now and then some blurred cliff or headland. We were glad to reach the open sea where, by the time we had made an offing of 11 miles, and taken our departure for Cape Farewell, some 300 miles to the north-east, the sky had cleared and the wind moderated; but astern Fiordland still lay buried in cloud and rain.

In that area where strong winds from a westerly quarter prevail, we had expected to make a fast passage, but in this we were disappointed, for apart from one wild night of squalls and rain, what little wind there was remained variable, and for two nights we lay becalmed, rolling heavily in a high swell from the north-west, which was probably caused by a tropical disturbance that had been battering the Queensland coast of Australia. Therefore the trip was slow; on one day we made a run of only 15 miles, and our best did not exceed 114.

During the night of our fifth day out from Milford we had Cape Farewell, from which Cook took his departure for Australia, abeam, and keeping a respectful distance off the shore, which from that point on is low and featureless, crept along with faint airs through the remainder of the night towards the powerful light on Farewell Spit, and daylight had come by the time we had it abeam. A nice breeze then made, and we went romping on our way round the spit and into Tasman Bay, a lovely little cruising area where we found good anchorage in a cove on the inside of Adele Island where there was another yacht, the first we had seen for some weeks. On the mainland shore a few small baches looked out from among the trees above a golden beach; as there was no road the only access to them was by boat.

In the morning we sailed across the bay to Nelson. Its harbour is formed by a natural and remarkable boulder bank 7 miles long, one of nature's curiosities, and as the navigable part of the harbour is small we wondered where to go as it is used by merchant ships and we thought we might be in their way if we anchored. However, at the main wharf we saw a man waving a red flag, and supposing divers to be at work, although we could see no diving boat, we stopped. The flag seemed to be beckoning, so we went cautiously towards it, and as soon as we were within earshot the flag-man told us to come on in round the end of the wharf as he had a berth for us there. It turned out that the yacht we had met the previous evening had radioed the harbourmaster on our behalf, and we were soon tied up comfortably alongside a suction dredger, where we had many visitors some of whom, in typical New Zealand fashion, took us to their homes for meals and drove us about the fertile plain to see the orchards and the tobacco fields; others were helpful with the small jobs that had to be attended to on board.

Everyone spoke so highly of the beauty of the Marlborough Sounds and the many wonderful anchorages to be found there, that we agreed we could not leave the South Island without having a look at them. To get there from Nelson it is usual to go through French Pass, a channel restricted by a reef to a width

▶

27. Winter quarters. *A*: Near the town of Whangarei (seen here from the top of 800-foot Parahaki), and 14 miles from the sea, *B*, we found a snug berth well-sheltered by muddy banks from every wind. *C*: Susan fires up the boiler to get a hot shower at the simple little marina where we lay. *D*: Nicholson uses his bird-watching verandah, a shelter I had to build over the sloping coaming aft to prevent rain falling straight through the opening port into the sleeping-cabin.

A ▲
C ▼
B ▲
D ▼
PLEASE
KEEP WOOD
AWAY FROM

of half a cable, separating D'Urville Island from the mainland; through it the tidal streams rush at 7 knots. A light fair wind and a fair tide carried us towards it, but as the tide gathered momentum and equalled the speed of the wind, we lost steerage way and therefore motored. A small buoy marking a middle ground flashed by, and almost at once we were in the narrows. To starboard on the steep-to shore of the mainland stood a lighthouse, and to port, at the end of the reef reaching out from D'Urville Island, stood a lighted beacon with the tide boiling past it and over the reef where it appeared to be running downhill. Moving as we probably were at 12 knots over the ground, we were through in a few moments and out on to the smooth water beyond. In the days of sail French Pass must have been difficult and often dangerous. Cook did not attempt it, but D'Urville did and nearly lost *Astrolabe* in the process; a lively painting by Le Breton, now in the Turnbull library at Wellington, depicts the incident.

We spent two weeks in the network of the Marlborough Sounds and found them to be very squally (only one of the few yachts we saw there had her mainsail bent on), and with their barren, yellow hills, ferries and holiday homes, they did seem a little disappointing after the wild majesty of the lonely fiords from which we had come. Then, on one of its rare gentle days, we crossed Cook Strait to Wellington, capital of the Dominion, and were given a berth in the marina at Evans Bay, where ours was the only inhabited yacht, for officially one was not permitted to live aboard in Wellington Harbour in case one might pollute it, yet the city sewers were discharging there and so were the cruise ships; indeed, we were advised not to seek a berth in the yacht basin off the Port Nicholson Yacht Club because of the sewage which drifted into it when a passenger liner was in port.

The objection to yachts pumping out their heads is becoming increasingly widespread, particularly in the U.S.A., yet it had been found that in Long Island Sound the many thousands of yachts that use it were causing a mere 0·07 per cent of the total pollution in that stretch of water, and that anyway their organic

◀

28. Rough weather in the Tasman Sea, and for the best part of three days we lay hove-to. But because Tasmania was not far to windward the sea did not grow high.

waste was converted by bacteria, fish, etc, into something else very quickly. New York, among some other states, insisted that all yachts must have holding tanks (it was even an offence to be sick over the side), but nothing like enough pumping out and disposal stations were provided. Now that so many people and bodies have become interested in the environment, they rightly feel they must make a start on pollution control somewhere, and the easiest target, for it is an almost helpless one, is the yacht. If only the environmentalists could have started at the other end, devoting their energies to curbing the actions of oil terminals, factories, commercial and naval ships, and city sewage outfalls, the end they sought might have been better served. Then when they had at last stopped the flow of that mass of largely inorganic filth, we could not have objected if they wanted to seal up our Baby Blakes.

Standing as it does beside Cook Strait, Wellington naturally has a reputation for frequent strong winds. We had one day of calm, and then a north gale blew with such fury that aircraft taking off from the near-by airport rose almost vertically, and we feared that *Wanderer* might pull the ring-bolts out of the marina; so, much to the amusement of the locals, who reckoned it was not blowing properly until it reached 80 knots, we unshackled our anchor and wrapped its chain right round the main walkway; this ensured that if we did move the whole marina would come with us. The meteorologists predicted that on the sixth day the wind would moderate and then swing round and blow from the south. That was just what we wanted, so when the wind began to moderate one evening we planned to take the last of it out of the harbour in the morning, and then hope that the wind change would come by the time we reached Cape Palliser, as our course would be north of east after that.

However, Nicholson had other ideas. While we were eating breakfast he nipped ashore, which of course he was forbidden to do while in New Zealand, and although we spent many hours looking and calling for him, he did not return until evening, when he sauntered on board and washed himself; by then the wind had changed, but it blew with such force from the south that there was no possibility of extracting ourselves from the marina for the next five days. However, people were very kind and hospitable, and one evening to their astonishment, for apparently they would not think of going there except by car, we walked to the summit of Mount Victoria to watch the lights come on in the hill-cradled city at our feet.

Our trip north was uneventful until we learnt that a cyclonic disturbance was moving south to meet us; this had already brought the usual gales and floods, with which such things are associated, to Northland, and was expected to produce storm force onshore winds in the East Cape area through which we had to pass. Therefore, instead of keeping the coast close aboard as we would otherwise have done, we steered to give East Cape a berth of 50 miles. That may seem excessive, but 18 miles east of the cape lies Ranfurly Bank; this carries a least known depth of 10 fathoms, but over it the tidal streams run at from 3 to 4 knots, and we thought a dangerous sea might well be found there in gale conditions. If we were to experience the forecast strength of wind (48 to 55 knots) we might have to run before it, so we needed to have ample searoom between us and the bank.

Having made our offing we hove-to, and remained so for the best part of two days. The wind certainly never reached storm force, but it was wild weather with much heavy rain and a high and angry grey-green sea, and we were thankful to be where we were instead of near the bank. The motion was violent and the noise disquieting, but at least we were dry down below, for we had no deck or hatch leaks, and since we had worked on the opening ports none of them dripped. When the glass started to rise and the weather to improve we sailed on, heading to the west now, and had a magnificent night sail under a full moon with a 30-knot wind on the beam, during which we passed through the lee of volcanic White Island where the smell of sulphur was strong enough to give us a good position line. So we came to Tauranga, and having decided which of the profusion of objects were navigational marks and which were local boats out for the Sunday afternoon, made our way up the harbour and secured alongside the public pier, which was separated from the attractive little town only by a railway line and a strip of well-kept garden ablaze with flowers, many of which we had not seen before.

As always in New Zealand, Tauranga showed an interest in voyaging yachts and their people, and at our berth, which was so easy of access, we almost constantly had a crowd of onlookers; during the day, and sometimes before we had got up in the morning or after we had turned in at night, we could hear quiet voices discussing *Wanderer* and her gear in a knowledgeable way, and it was rarely possible to run the gauntlet from saloon to the after heads without being addressed:

'What happens when that cockpit fills up?'—'Why did you choose ketch rig?'—'What varnish do you use?'—'Just the two of you, eh? My, but she's a big 'un for two!'—'Excuse me, but have you sailed all the way from Auckland?'

Questions addressed to Susan when she ventured on deck tended to be more psychological:

'How did your old man get you to go to sea? My missus won't look at it.'—'Don't you ever get frightened?' or 'tired?' or 'seasick?' And often someone would hand her a gift and say: 'Here's a snapper for your tea,' or it might be: 'Just a few roses from the garden.'

It was fun; we enjoyed it all immensely for a week, and then continued our trip to the north. Again we visited Auckland and pottered about in the Hauraki Gulf, and we spent a wonderful, silent week in perfect autumnal weather out at the Great Barrier Island, where we had the good fortune to meet 'Stud' Stellin. His career had been remarkable in that he had served in the navy, the army, and the air force, and now was trying to farm a large section on the island. After his house had been burnt down he built another and a better one, but his wife decided to live in Auckland and breed Siamese cats. Once a year Stud used to visit her and stay three months. He had a very small launch in which to make the 50-mile trip, and as he had a pet goat of which he was fond, that had to go along, too, rather, we gathered, to his wife's dismay, for it expected to travel in her car and live in her garden.

Finally, and shortly before midwinter, we tucked ourselves up to wait for the spring in Alan Orams's little marina 14 miles inland and only 20 minutes walk from the Northland town of Whangarei (Plate 27), where we did a lot of work on board. The marina was a simple affair, though comfortable; there were no carpets, no rubber-tyred carts, no electricity, but arrangements could be made for a paper to be delivered each morning; if an empty milk bottle with 4 cents in it was placed by the roadside in the evening it would be replaced with a full one overnight (we never once saw the milkman); water was laid on to the pontoons, garbage was collected, and there was a lean-to shower stall beside the boatbuilding shop. To obtain a hot shower one collected a supply of wood, and if this was wet, a can of turpentine, and lit a fire under an outdoor boiler, which once had been lagged, but now looked a bit moth-eaten. Twenty minutes later, and it was not wise to wait longer or the shower might be a steam one, the water would be hot and plentiful, always supposing no shipwright had altered the cocks on the boiler so as to steam-bend some piece of timber.

We might have enjoyed our stay at Whangarei much more if we had not there lost our gay little companion of the past five-and-a-half years. Nicholson became ill with an incurable kidney complaint, and we had to have him put away. He died in Susan's arms, and the ship thereafter was sadly silent and empty without him. No matter how frightening conditions afloat or ashore might have been, he always kept his flag—his thin, black tail with the kink in it—flying.

# 8

# Heavy Weather in the Tasman Sea

Susan and I consider that Whangaroa in the far north is New Zealand's best and one of her most attractive harbours. Its entrance, which is little more than a cable wide and, is protected to some extent by off-lying Stephenson Island, is easy and free of dangers, and the swell dies as soon as one is within it; the scenery is spectacular and the anchorages are excellent. It was to this lovely place that we came at the end of our year in New Zealand to wait for reasonable weather in which to round North Cape and start our crossing of the Tasman Sea. But we found we could not relax and enjoy it properly because we were anxious to begin what we thought was likely to be a difficult passage; but day after day the weather remained unsettled and strong headwinds blew to keep us in port.

Well up the harbour and standing on opposite sides of it were two tiny villages, each with a small store from which most simple provisions could be had, and one of them, which was built on piles in the water, had a hose from which fresh water could be taken straight into the dinghy. This was of particular interest to us, for drinking water on board had become something of a problem. During the winter we had decided to open up and cement-wash the insides of the tanks, for as their lower and outer sides are formed by the skin plating, we could not afford to have any rusting or corrosion there. This operation involved removing the Kempsafe heater and the saloon sole, and unbolting four manhole covers, each of which was held by 60 rusting studs, some of which broke off and had to be drilled out. When news of what we were up to got around, a professional plasterer arrived with a gift bag of cement and instructions on its use, and an expert steel-worker lent us the tools needed for dovetailing neoprene gaskets and punching each of the many stud holes in the right place. This was typical of the help that was so frequently offered us; even on the rare occasions when I tried to get a professional to do a job for me, he usually produced the tools and the materials, showed me how to use them, and said: 'You don't want to waste your money paying me to do this.' As Susan is more lissom than I am, she inserted herself into each tank in turn to clean it and paint on the cement, and there were times when I feared I might have to use a hacksaw to release her. For a week we lived a spartan existence on the cold tank-tops, avoiding their gaping holes which exuded the alkaline smell of wet cement, and it was a great day when we got the covers bolted on and the sole and heater replaced. But from that time on, although the tanks had been repeatedly filled and emptied, the water came out tainted and undrinkable, for Whangarei water was heavily chlorinated, and it seemed that the chlorine was reacting with the fresh cement. For the Tasman crossing we therefore had to rely for drinking on water carried in plastic containers.

On the sixth day, after Susan had replenished our supply of bread, butter and eggs, we sailed away, but a few hours later the conditions became so unfavourable for rounding North Cape that we bore away and anchored in Houhora Bay at the foot of Mount Camel, so named by Cook 'because it stands upon a desert shore which is in general low, and the soil to all appearances nothing but white sand thrown up in low irregular hills.' Since our trip round the South Island we had replaced the 60-lb CQR bower anchor with one of 75 lb, and that night it had its first real test, for shortly before midnight a gale sprang up from out of the north, and although the horn of the bay just sheltered us from the sea, the wind came over it in violent squalls, and uncomfortably close astern the swell broke on the other rock-fringed horn of the bay. We kept watch for four hours, after which the wind moderated—and the new anchor held.

The morning forecast was optimistic, but we did not like the look of things at all: the swell was higher than usual and was coming from north-east; there was a solar halo, the glass was falling, and haze soon came to take all the sparkle out of the scene. These portents suggested we might be wise to leave our anchorage, which was open from nor'-nor'-east through east to south, and return to the security of Whangaroa 35 miles away. We did so, and as we approached the narrow slit in the coast that is the entrance, the visibility was growing poor, and the swell was breaking sullenly at the feet of the high, black cliffs.

Not long after we had reached our chosen anchorage, the wind started to come in gusts, each stronger than the last, and by the time our conservative hand-held anemometer registered 44 knots on deck, the new anchor was having another good test. The gale continued for 24 hours, and we learnt that Cape Reinga lighthouse, which we would have been trying to round had we acted on the weather forecast and ignored our own observations, recorded the wind at 81 knots, which is high up the hurricane scale.

A few days later the weather showed an improvement, and we slipped out of our haven, which was looking particularly lovely with its cliffs and rock spires bathed in the low, warm, early morning sunlight, and headed for North Cape which we reached in the afternoon; but there we lost the wind and motored on through huge shoals of fish about which the sea-birds were busy, and by evening we had passed Cape Reinga and were out in the Tasman Sea.

The pilot charts for the South Pacific show that there is no month when the Tasman is free from a risk of storms; on the charts their scarlet lines criss-cross it throughout the year, but at least December, when we were attempting our crossing, is not so bad in this respect as are some other months. By 'crossing the Tasman' most sailing people mean making the passage between North Cape and Sydney in New South Wales; but that was not now our intention, for we had done it on an earlier voyage, and this time our hope was to make good a south-westerly course and sail direct for Hobart in Tasmania in 43°S. latitude and distant about 1,400 miles.

That night we picked up a fair breeze and had pleasant sailing for the next two days, mostly with a running sail set (we call it that because it is cut flat and is quite unlike a spinnaker). Originally, as I have mentioned earlier, we had a pair of such sails for use when running in the trade winds, but did not find the arrangement very satisfactory. So while in New Zealand we had abandoned the rig and had a fitting made to take a single boom low down on the fore side of the mast, the boom when not in use being stowed up-and-down the mast where it was held by its own topping lift. Thus we could lower it for use on either side, and hold it in position fully under control while the sail was being outhauled and set. This was the first occasion we had used the new arrangement, and it worked well, but when on the third day out the wind increased to 30 knots and the running sail had to come in, we found that the strain, which must have been considerable, had twisted the new stainless steel mast fitting so badly that for the rest of the passage we did not care to use it again for fear it might carry away and let the boom run amok.

By that time we had already reefed the mainsail and handed the mizzen; but the wind and sea were increasing steadily and the auto-helmsman could no longer manage; so we took over the helm, but after some hours grew so tired that we hove-to for the night. By morning the wind seemed to be too much even for the close-reefed mainsail, under which the ship was staggering in the squalls; so we took the sail in, and with some difficulty because once we had started the halyard the sail beat wildly at us as though determined to knock us from our precarious foot-hold on the coachroof. Normally 5 tiers keep the stowed sail neatly in place, but with so much wind that was not enough, so we lashed it from end to end with rope. We then lay a-hull, but as the hours passed and the sea continued to build up, I began to question the wisdom of remaining beam on to it (I had but recently been reading Miles Smeeton's strictures on those who advocate lying a-hull) and felt there was some risk of a breaking sea filling our large, centre cockpit, which by my reckoning would hold about 4 tons; although we had by now shipped the washboards in the companionways, and the strongbacks which I had recently made for the saloon and cabin doors, I feared that if we did ship a heavy crest amidships it might burst in the doors and flood the accommodation. In the comparative peace of the saloon, where the gale could not tear at us, and its deep, vibrant note was muted a little, we discussed the situation, which was one we had not met before in this *Wanderer*, and agreed that the only remaining course of action was to run before it. So we went out into the cockpit again, where the wind pressed our oilskins tightly against us and whipped their collars and skirts, unlashed the wheel and put the helm up so as to bear right away and run before the wind and sea. It was the biggest sea that *Wanderer IV* had ever seen, and it certainly was impressive. Under bare poles she sailed at 4½ knots. There were moments when with her stern flung up high she seemed to be poised at an alarmingly steep angle as though preparing to take a header into the trough which had just passed and now momentarily lay so enticingly ahead of her gold-tipped bowsprit, and we felt we needed to steer with care, always keeping the wind on the back of our necks, if the next rapidly approaching crest was not to catch us on the quarter. It was all very well, I thought, for Bernard Moitessier to state that the heavier seas *should* be taken on the quarter; he was not with us, and I was not prepared to experiment in the dangerous conditions which now prevailed, and I felt that our safety lay in presenting the smallest possible target to the monsters coming up astern. One needed to concentrate and to turn the wheel smartly to achieve this, and it was tiring mentally rather than physically, and I did wonder how long we could keep efficiently at it; but how thankful we were that before meeting the Tasman at its toughest we had cured our ship's steering vices, otherwise the helmsman's job would not have been possible. Fortunately for us her stern continued to lift buoyantly to every overtaking sea, and she shipped nothing but some heavy dollops of spray and the rain which continued unceasingly for 48 hours. I do not know for how long we continued to steer, taking spells at the helm of one hour each, for I omitted to make a note in my journal, but it was a tremendous relief when at long last the note of the gale in the rigging dropped an octave, and a few hours later the sea, although still running very high, no longer looked quite so steep or had such dangerous-looking crests, and we felt it would be safe to let the ship lie a-hull once again and look after herself while we changed into dry clothes, had some food and slept.

That violent gale pushed us some way north of the rhumb line, but throughout the next six days we were able to steer in the right direction mostly in fine weather though often with insufficient wind, and to save from chafe the new mainsail which we had recently bent on, we took it in and sailed on under the curious but effective rig of jib, mizzen, and mizzen-staysail; but the latter had to be tacked down to leeward instead of to windward because the mainsail stowed on its boom was in the way, and the sheet of the high-cut No. 2 jib therefore girted its foot. At times we regretted being unable to use the running sail, but perhaps that was just as well, for one morning something clicked in my left elbow, and for the rest of the trip I was unable to make much use of that rather painful arm; so Susan took over all the sail handling, which normally is a joint job or mine alone, and did it in her usual quiet and efficient way, but I hated watching her from my safe place at the wheel as she worked on the reeling foredeck; in the past she had bad trouble with her back, and I feared that so much hard work might cause that to occur again.

On the tenth day out of Whangaroa we had come into a 5° square on the December pilot chart in which on one day in ten a gale could be expected, and in which the longest wind arrow flew from the north and had six feathers on it, indicating an average strength of Beaufort force 6, i.e. 22 to 27 knots. It looked as though we were not going to get across that square without some unpleasantness, for the weather portents had been bad for the past three days, with solar halos, a torn sky, and a falling barometer, and so great was the humidity that four hours before sunset the sails and deck were dripping; meanwhile the misty sun shone palely like an indistinct orb of silver, just as we had seen it before a sandstorm in the Red Sea. These omens proved right, and the next day found us lying a-hull with a west-nor'-west gale of 60 knots playing its high and low organ notes in the rigging, and whitening the sea with crests and long streaks of spume. But as the nearest part of Tasmania was to windward and only about 160 miles distant, the sea, though rough indeed, did not reach the height it had achieved in the earlier gale and did not constitute a danger.

During one of my periodic tours of inspection to see that all was well on deck and below I discovered two things which were rather alarming. There was some water sloshing about in the usually quite dry bilge of the engine-room, and I traced its source to a leak in the exhaust system of the auxiliary generating plant. The outlet at the ship's side had a short stand-pipe with a gate valve on it, and although that valve was screwed down tight it was obviously letting water back into the exhaust pipe and twin silencers; somewhere the system must have been broken or holed, for the water was percolating through the asbestos lagging. At sea the only possible cure was to go over the side and drive a bung into the stand-pipe, but in the present weather conditions that was clearly out of the question; so I pumped out the bilge and continued to do so thereafter every few hours for the rest of the trip, and hoped that meanwhile none of the water was getting into the engine. (Eventually we had to renew the entire exhaust system.) The boom gallows, on which the heavy main boom with the stowed sail on it was resting, consisted of a teak crossbar supported at each end by vertical stainless steel tubes, the feet of which were welded to plates on deck. I found that the welds were giving way, so that each time we lurched there was a slight athwartships movement of the gallows. I put on preventer lashings and prayed that the gallows would not collapse under the load imposed by the wild motion. While I was on deck I re-lashed the bower anchor, which had worked loose, and gathered in two of the sheets which had been washed overboard.

For most of three days we lay stopped, either a-hull or hove-to under short sail, and although this was frustrating when we were so near our objective it was no great hardship while the heavy rain, which accompanied the early part of the gale, lasted; but when the sky cleared to a cloudless pale blue we grew impatient, and although Hobart and Melbourne radio stations continued for three days to broadcast gale warnings, and told us of damage done ashore and at sea, we felt impelled to move on, for our intended landfall, Tasman Island at the south-east corner of Tasmania, lay only 150 miles away. So

▶

29. *Top*: In Horseshoe Inlet, Port Davey, we found a perfect anchorage among hills not too steep or high to climb for the benefit, *bottom*, of the views they afforded; looking east through Port Davey the neck that carries the South Coast Track can be seen in the middle distance.

MF

Susan set the staysail and the mizzen and steered while I set about getting breakfast of boiled eggs and toast. But now that we were moving, and quite fast, the motion was more violent, and one especially sudden leeward lurch tilted the swinging cooker to such a steep angle that it jammed—it has to go beyond 50° for that to happen—the eggs were tossed on to the floor, where they naturally burst, some of their boiling water managed to reach and soak the buttered toast, which was in the warming cupboard under the hotplate, and then, remarkably, the sodden toast was flung up on to the saucepan rack—and all this happened in a few seconds. So we had coffee and biscuits instead, and we persisted in sailing for a few hours, but by then had had enough, so we hove-to again under the small area of sail that happened to be set.

In the morning of the fourth day of that long-lasting gale, we discussed from the saloon settees, on which we lay, held by canvas bunkboards, whether we should now abandon our attempt to reach Tasmania and head instead for Sydney and quieter weather. Susan is usually the stronger spirit when decisions of this sort have to be made, but oddly enough this time it was I who argued in favour of continuing for our chosen destination—when we could. And Providence must, for once, have been on my side, for although Hobart radio was still continuing to talk about force 9 gales, the wind began to moderate so that within a few hours Susan was able to get some more sail up, and we proceeded bumpily on our way. The wind continued to take off, and so quickly that Susan was hard put to it to keep pace making sail and—all or nothing seems to be the motto of the Tasman—very soon we were motoring with no wind at all across a steep and very confused sea, and shortly before dawn of our fifteenth day out we sighted ahead the brilliant flash of the light on Tasman Island.

We could scarcely have wished for a more satisfying and spectacular landfall than was revealed to us as the grey light of

◀

30. *Top*: Denny King and his artist daughters, Mary and Janet, in their lonely house, which Denny built, in the Port Davey wilderness. *Bottom*: 'Let Bob do the worrying'. Not often does a bulldozer come to the assistance of an aeroplane stranded on an uninhabited island.

dawn spread over the scene. The 800-foot island is separated from Cape Pillar by a quarter-mile-wide channel, and the architecture of both island and mainland is remarkable: great flat slabs of rock, which look like man-made forts, are supported by an army of perpendicular pillars of basalt.

During the final voyage that we made in *Wanderer III* we had, while at Bequia in the West Indies, met the Cuthbertson family taking their Scottish-built vessel *Kathleen del Mar* out to Tasmania to fish for crays. Since then we had corresponded occasionally, and the last letter we had told us that they now lived near Hobart. Even so it was a strange coincidence that as we came to the commercial port at Hobart needing customs, health and immigration, and uncertain where to berth, Bern Cuthbertson should be standing on one of the wharfs beckoning us in. He quickly had the swing-bridge opened to let us through, moved a few small craft to give us room, and very soon we were tied up alongside the wall in a corner of clean little Victoria Dock in perfect security and right in the heart of the town, where the kindly officials, who had not entered a British yacht in all the years they could remember, for most foreign yachts call first at Sydney, quickly did what was necessary.

That evening we were driven across graceful Tasman Bridge in Bern's big car to dine luxuriously with him and his family in their fine old Lindisfarne home. Bern, we learnt, had abandoned crays in favour of the more numerous and more profitable abalone, and for that purpose had bought an old 110-foot river steamer into which he had put a Gardner diesel, and with four divers was working the abalone grounds off the west coast. Somehow we had hardly expected our fisherman friend to live in a manor house, have his name on his wine bottle labels, or to have two vintage cars in his garage; neither had we thought to find the bulwarks of his creaking old ship carrying in huge yellow letters the slogan: STOP POLLUTING OUR SEA. But Bern is a Tasmanian, and like Tasmania is remarkable for his contrasts, surprises and character.

Whenever it is convenient I like to go for a walk before breakfast, and our berth in Victoria Dock was particularly suitable for this. Within a few minutes of stepping ashore I could be walking

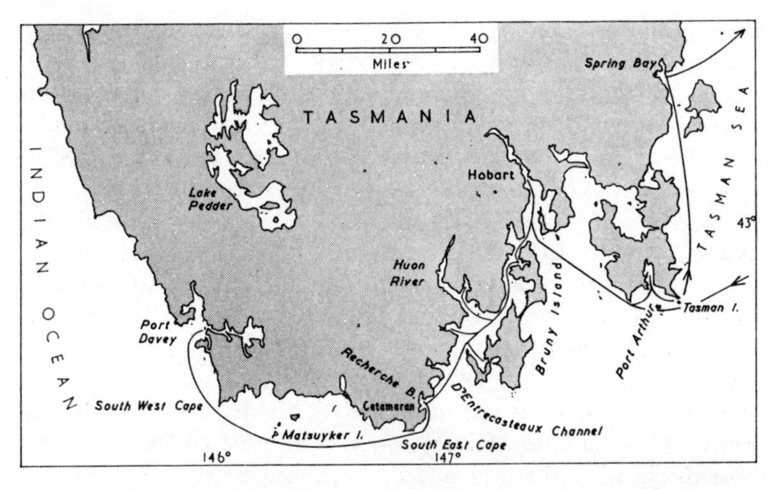


past sleeping St. David's Cathedral and up Elizabeth Street, which at my early hour would be almost devoid of traffic, or exploring Salamanca Place where still stood the warehouses used in whaling days, or approaching Parliament House across the springy turf of its tree-shaded park. I found that Hobart had a character peculiarly her own, and was remarkable for her clean, sandstone buildings, even though some of them might be roofed with corrugated-iron. But it did seem that there had been too little check on the erection of high-rise blocks of glass and concrete; low and graceful Parliament House, for example, was dominated by a skyscraper, and the charming, pale green, two-storey Marine Board office was being replaced by an austere, grey monster right alongside.

At the dock many friendly people stopped to talk and offer us hospitality, and we found them relaxed, cheerful, and very proud of the island state in which they lived. One of them drove us out and showed us round the vast hydro-electric scheme in the west, which was causing wide concern because of the imminent flooding of Lake Pedder, another had our boom gallows repaired, and the bent fitting for the running sail boom straightened and strengthened, and nearly everyone said 'You must go round to Port Davey, it's quite out of this world.' But when we produced the charts which had been lent to us and asked which anchorages they recommended, it turned out that hardly any of them had ever been there by sea. However, when we learnt that only two families lived on the shores of that remote harbour which thrusts into Tasmania's wild, west coast, and that the nearest road ends 40 miles from it, we agreed we must certainly go there, so early on Christmas eve we had the swing-bridge opened to let us out, and sailed down the Derwent River.

Twelve miles south of Hobart lies the north end of Bruny Island, named after the Frenchman Bruny D'Entrecasteaux, who explored those waters with the ships *Recherche* and *Esperance* in 1792. The north part of Bruny is joined to the south part by a long, narrow isthmus, the whole extending for 27 miles and giving good shelter to D'Entrecasteaux Channel, which separates the island from the mainland. This stretch of water, together with the Huon River, abounds in good anchorages, and provides a pleasant and attractive cruising ground, which is flanked on its western side by a range of mountains. But everywhere in that area we saw the stark, white trunks of the trees killed by the great fire which had raged over it seven years before. Even Bruny Island did not escape the conflagration although the channel at its narrowest is three-quarters of a mile wide, for the fire coincided with a westerly gale which carried incendiaries across nature's firebreak.

We spent a pleasant fortnight cruising in the area, which is rich with names nostalgic to anyone from the U.K.: Kettering, from which the ferry runs to Bruny Island; Margate, with its carbide factory; Lymington, where sleek cattle grazed, and apple orchards patterned the hillsides; Dover, a tiny fishing port; Strathblane, a timber mill to which the last survivor of the coastal trading ketches, the *May Queen*, was still going to load planks for Hobart; Folkestone, where there was nothing much except an excellent anchorage behind a small island ruled by a goat which fiercely repelled boarders.

D'Entrecasteaux Channel is remarkably free of dangers, but after we had left its placid waters astern and were nearing

▶

31. *Top*: After looking at the chart I thought that Lindisfarne Bay offered the best anchorage in the neighbourhood of Hobart. Tasman Bridge, under which we had come to reach it, lies in the background lit by the rising sun. *Bottom*: '. . . the misery, the utter hopelessness . . .' The ruined buildings of the convict settlement cluster round the huge penitentiary block at Port Arthur.

Recherche Bay, pilotage had to be taken more seriously, for the approaches to that bay are beset with several reefs, most of which are submerged and are marked only by occasional breakers or perhaps by kelp. However, there are large-scale charts of these waters, and we had no difficulty in reaching the Pigsties, which offers the last anchorage that side of Port Davey.

Having visited Oban, the southernmost settlement in New Zealand, we felt it would be appropriate to have a look at Catamaran, the southernmost settlement in Australia where, the *Pilot* informed us, coal is mined. So we made our way south for several miles along a rough road, but although we must have passed right over the site of Catamaran we never saw a sign of it. Eventually the road ended at a footpath, which was the beginning of the South Coast Track leading to Port Davey, a walk of some 40 miles, which the active can manage in 2 days, and the less active in 6; as there was no road to Port Davey, either one had to go on foot or by sea (by the latter the distance is 80 miles) or, if a tourist, possibly by light plane. The following day we walked north to have a look at Leprena, which we supposed must now be Australia's southernmost settlement, but again could find no trace of it; however, in that bush country wooden buildings do not survive unattended for many years; coal is no longer mined, and felled timber is transported by road.

For some time I had been concerned at the trend of the modern British chart. In common with the majority of English-speaking seamen, I dislike the idea of having to change my thinking from fathoms to metres, but more serious to me is the fact that the chart tends to give less and less information. I know it is expecting too much to find on the modern chart the little gems of local knowledge such as the older ones often carried. I well remember one which read: 'There is a considerable indraught into all the deep bights between Portland and the Owers, particularly on the flood round Durlstone Head into Poole Bay.' However, soon after leaving Hobart I found a typical example of my chief complaint: the practice of adding

32. *Top*: The Bay of Islands on the north-east coast of New Zealand. We entered it after nine days of almost continuous hard headwinds, and *bottom*, anchored off the little town of Russell, which was circled by a rainbow.

unnecessarily to the navigator's work. On the Hobart to Norfolk Bay chart there is, near Betsey Island, a note which reads: 'Local magnetic anomaly, see Caution No. 2.' Caution No. 2 read: 'A magnetic anomaly is reported to exist in the area indicated on this chart. See Admiralty Sailing Directions.' So I referred to the *Pilot*, which took a little time because magnetic anomalies are not indexed, and read: 'A local magnetic anomaly, affecting the compass to an extent of about 13° in either direction, is experienced in an area west of Betsey Island.' Approximate boundary lines were then given. This small but important item of information, for which the navigator has to spend time searching, could so easily and clearly have been put on or near the area affected on the chart, which had plenty of blank space suitable for the purpose. I learnt from a report in *The Journal of the Institute of Navigation* (now *The Journal of Navigation*) that those lovely little sketches of the coast, which decorated the borders of many charts and have so often solved problems of recognition for the stranger, are to be abolished, the argument being that today all vessels of any importance have radar, and that anyway the sketches get out of date; they do indeed, but I imagine an artist like David Cobb would love to be given the job of updating them. I learnt, too, that isogonals (lines of equal magnetic variation) are likely to be deleted from ocean charts, so that those of us who use the magnetic compass and make a course alteration each time we cross one of those lines, will in future have to consult some other publication. Magnetic compass roses are rapidly being removed as new editions are published, so that a small sum has to be worked when laying down a course or bearing, and in the near future it may not even be possible to tell from the chart whether the beach is of sand, mud, stones, or rock.

While we were at the Pigsties we had the good fortune to meet Admiral Sir Guy Wyatt in his handsome ketch *Saona*. As he had been the Hydrographer of the British Navy, I was particularly interested to discover that his thoughts on the modern chart were in close agreement with my own, and that some of the changes appeared to have been made more for the benefit of the cartographer than the convenience of the navigator. Sir Guy had a great love and understanding of his subject, and since his retirement to Tasmania he had personally surveyed a number of bays and inlets which were not covered by the official publications, and he kindly provided us with prints of some that he had done of parts of Port Davey.

On our trip round to that place there was so little wind that we did not reach South West Cape until late afternoon. The course from there to Port Davey was nor'-nor'-west for 15 miles, give or take a mile or two, for some of the islands and headlands along that coast were known to be incorrectly charted, as we were to prove to ourselves later. This distance, together with the run in to an anchorage, was more than we could accomplish before nightfall, and as we did not feel we could enter safely in the dark we hove-to on the offshore tack.

The night was magnificent, clear and moonlit, but although this was midsummer it was so cold that we wore two sweaters even when in our bunks, and the great light on Matsuyker Island, south of which we had passed that afternoon, gave its group flash every 10 seconds to keep us company and provide a check on our latitude. At midnight we wore round on to the other tack, and at dawn found ourselves in the same position that we had occupied 8 hours earlier. We then made our slow way up along the coast, seeing little of it except its outline because the sun was in our eyes. However, with the help of some off-lying islands we identified the entrance to Port Davey, and stood in towards its entrance, which is well protected by the thin wall of the Breaksea Islands, south of which we passed. In the approach to that passage there is a rock with less than 6 feet over it, so we checked our position and our progress carefully by cross bearings.

Port Davey, with Bathurst Harbour into which it leads, includes a stretch of sheltered water running 13 miles in an east-west direction (Plate 29, *bottom*), and there are two large-scale charts of it; but detailed though these are, the approach is rather sketchy, with dashed lines in places to indicate that there the coast had not been surveyed. So perhaps I should not have put too much faith in my bearings, though we got in without any trouble; but when on leaving 9 days later I laid down a transit of a charted headland and near-by island, I found that it placed us more than a mile from where it was obvious we really were.

As soon as we got in behind the Breaksea Islands we lost the swell, and on rounding a headland the fine harbour opened up ahead. It thrusts in among the mountains, which rise to 2,000 feet and more, and they were not tree-clad, but mostly were covered with coarse grass, scrub, and a multitude of wild flowers with stems so short that they hugged the ground to protect themselves from the great westerly gales; but a big area on the northern side, including the flank of Mount Rugby, was of a rust-brown colour due to a fire which had recently swept over it. At its narrowest point where a tongue of land reaches out from the southern shore, the harbour is only 200 feet wide, and it was there that a boat was provided at each side so that bush-walkers following the South Coast Track might cross over, three boat trips being needed so that a boat might be left at each side for the use of the next party coming from either direction.

The harbour abounds in anchorages, and for our first we chose one alongside the narrow tongue of land carrying the track, and by a strange coincidence, for bush-walkers are not all that common, were able to watch two parties making their laborious way along it, well booted against the risk of snake-bites. We found good holding in 2 fathoms, and were glad of it when the afternoon wind came roaring down on us from off the high land.

In most of Port Davey the water has a peculiar coffee colour, which entirely hides any signs of dangers or uneven bottom, and is caused by the freshwater streams which flow through the peaty soil of the button-grass plains in their course to the sea. The colour is noticeable even in the smallest mountain stream. A remarkable feature of this water is that it becomes streaked with long-lasting foam on the slightest disturbance, just as though mixed with detergent, so that when moving through it *Wanderer* left astern two broad ribbons of snowy-white.

In the whole of this unspoilt wilderness there lived, as we had been told, only two families, the Kings and the Claytons, and both homes were on the shore of Melaleuca Inlet farther up the harbour. The pioneer King family had operated a tin mine for nearly 40 years, and Denny King transported the ore in 100-lb. bags round to Hobart in his yacht *Melaleuca*. Since Mrs. King's death he had lived there alone except for visits by his two daughters, who were at university. Clyde Clayton and his wife Winsome (she was Denny's sister) used to fish for crays in their old river steamer, but now were more interested in prospecting, and had already by hand mined a ton of some mineral for which they had so far failed to find a market.

We decided to call on these isolated people, but with the reservation that we would make our visits brief if it seemed that they did not welcome the intrusion of strangers, for we felt that the privacy of anyone living in such isolation should be respected. We therefore went round to Melaleuca Inlet and anchored in its mouth off the Clayton house. Immediately friendly Clyde and Win came off to introduce themselves and invite us ashore later. We then put the outboard motor on the dinghy and headed inland for the King house. The channel, which was marked with sticks, carried about 6 feet, and it wound in a delightful manner so that one could never see far ahead, and each point rounded revealed a fresh vista of bush-clad banks backed by mountains; at its inner end, after coming about 3 miles, we found ourselves in a wide and shallow lagoon, and across this and up yet another creek stood the house. As we approached the landing place, Denny (Clyde must have told him by radio that we were on our way) came to the water's edge to give us a charming welcome and lead us up to his home, which of course he had built with his own hands. There we met his daughters Mary and Janet who were busy with their brushes and canvases, for they were both artists and had recently had an exhibition of their oil paintings in Hobart and now were busy working off their commissions (Plate 30, *top*).

One might have thought that to live alone for most of the year, mine tin and transport it single-handed for 130 miles by sea round the stormy side of Tasmania, would be something of a problem, but Denny appeared to take this in his stride; his only problem was that he had too many visitors in the summer. Not long before he had featured on a television programme, his home and his mine were shown on a new, large-scale map, and for his own convenience he had made a small air-strip. So inevitably tourists came, and to Denny they were not quite the same as the people who walked or sailed round to see him. 'They all ask the same questions,' he told us, 'take the same photographs, and generally sabotage my day.' Nevertheless he smiled as he said it, and we believed that he enjoyed the sociality.

We found that Mary and Janet were not only good artists and naturalists but cooked extremely well, and when we left on the dinghy trip back to *Wanderer* we carried with us the best and most crisply crusted loaf that we had seen in many years. That evening by high-speed outboards the Kings and the Claytons, who came with gifts of home-grown fruit, joined us on board for drinks. We felt that it had been a great privilege to meet such remarkably self-reliant, hard-working and charming people.

As Susan had some laundry and I some darkroom work to do we moved round to Horseshoe Inlet (Plate 29, *top*), of which Sir Guy had given us a copy of his chart. It was a place very much to our liking; completely landlocked, so small that no wind could raise more than a tiny ripple on the dark water, surrounded by jolly little hills which were not too steep or too high to climb for the benefit of the view they afforded, and with a lovely sparkling mountain stream chuckling in a corner. Having avoided the 4-foot rock which lies invisibly in the entrance, we anchored and took a stern-line to a tree, and remained there for several days in great content.

It was there that for the first time we saw black swans, which are a little smaller than the British white swan, and make a different, less laborious sound while flying. One evening as the wind was dying away, a flight of 16 of these graceful birds, flying in a wide V formation with their long necks outstretched, came in over our mastheads and disappeared in the upstream lagoon. Apparently they did not find what they wanted, for within a few minutes the whole splendid flight sailed back with creaking wings and passed out of sight round the point under which we lay.

With our work finished we unmoored and went down harbour to spend our last night in Schooner Cove near the entrance, and shared that anchorage with two fishing vessels and two yachts, all waiting for the strong wind and heavy swell to moderate. We learnt that there were all told 17 fishing vessels waiting in the Port Davey area, for the cray-fishermen cannot work their pots nor the abalone men dive when a heavy swell is running.

We left as soon as it was light enough to see our way out. At sea we found a glassy calm, but as steep and confused a swell as we had met for a long time, and I had to abandon my attempts to get breakfast until we had rounded South West Cape and were running east, when the westerly swell, no longer rebounding off the cliffs, became more regular, and some wind in our sails steadied us a little. In the afternoon we rounded South East Cape, and in smooth water had a pleasant sail among the reefs off Recherche Bay and up into the now familiar D'Entrecasteaux Channel, to find a deserted anchorage in Bruny Island where we could listen contentedly to the roar of surf on the other side of the narrow strip of land that sheltered us from the Southern Ocean.

A few days later we returned to Hobart and picked up a mooring which the Royal Yacht Club of Tasmania had kindly placed at our disposal off its fine club-house in Sandy Bay. But we were too long for the space available, and when the afternoon onshore wind arrived we were at once in collision with a neighbouring yacht whose owner was fortunately on board. We slipped immediately, and not wishing to enter Victoria Dock with so strong a wind, passed under Tasman bridge, where there was so much sea that the hurriedly abandoned lunch things were rolled off the saloon table, and anchored above it in Lindisfarne Bay. From the chart I had assumed this to be the best anchorage in the vicinity of Hobart, and so it proved to be (Plate 31, *top*), and the people who lived in that suburb went out of their way to be kind and helpful. They seemed to be particularly upset when our dinghy complete with oars and outboard motor (we had been using the motor because my damaged elbow did not take kindly to rowing) was stolen from alongside one night while we were asleep. One friend lent us a dinghy, another went in search of ours and found her 3 miles away, and within 24 hours the alert police had apprehended two youths, recovered our outboard, oars and rowlocks from them and returned them to us.

Susan and I had been reading Marcus Clarke's long novel *His Natural Life*, a moving indictment of the transportation system to which many thousands of people were condemned in the early part of the last century, often for minor offences which today would merit no more than a small fine. The best preserved of the Tasmanian penal settlements to which those unfortunate people were sent stood on the shore of Port Arthur, a well-sheltered harbour near Tasman Island; so when we left Hobart to start on our way north, we put in there for several days to have a look at it, and anchored just off the cove round which the old prison buildings were clustered. The place was something of a tourist attraction, but by landing early we were able to explore it entirely on our own. Apart from its ease of access by sea and the ease with which the narrow isthmus could be guarded, the chief reason why Port Arthur was chosen as the site was that the main occupation of the convicts was felling trees and sawing them into planks, and that area had much fine timber growing on it. Although the yellow sandstone buildings, largest of which was the huge five-storey penitentiary, stood in a peaceful setting among tree-shaded lawns overlooked by the ruins of the chapel on its hill (Plate 31, *bottom*), not much imagination was needed after one had seen the barred windows, and the 'model' prison where complete silence was enforced, to picture the scene as it must have been over a hundred years ago, crowded with prisoners, many in chains, driven by overseers with whips—the degradation, the pain and the misery, the utter hopelessness. We came away feeling subdued and sad.

Our plan when we left Spring Bay, which appears to be the only good anchorage on the east coast of Tasmania, was to sail through Bass Strait to Port Phillip Bay, to which not many overseas yachts go, see some friends at Melbourne, which stands on the shore of that bay, and then cruise east and north along the Australian coast as far as Sydney before starting the voyage back to New Zealand. But with its strong tides and mostly unlighted islands, Bass Strait can be an awkward place at the best of times, and we had not got far on our way before I was again having trouble with my elbows, which made working ship difficult. So when we heard on the radio that an easterly gale was expected, it did not take long for us to decide that it might be more sensible to leave Port Phillip for some future occasion, and to head instead for Sydney, possibly by way of Eden and other east coast harbours.

In order to keep up to windward while we could, give Flinders Island and its off-liers a wide berth, and perhaps even get clear of the south-setting current, we steered a course 20° to the east of the direct one. But our observed position next day at noon showed that in spite of this precaution the current had set us some distance to the west of the rhumb line, and we could then not alter course any more to the east as the wind had backed a little and we were already closehauled.

In the forenoon of the second day we sighted land ahead and on the lee bow, and an observation of the sun showed that we had lost still more ground to the current and were now 50 miles to leeward of Gabo Island, which lies at the south-east corner of mainland Australia. We stood on towards the shore, hard pressed now and slamming badly in the head-sea although we had taken in the mizzen. One could not say that visibility was poor, for it was possible to see for 10 miles, but there was the same sort of haze as we had often experienced in strong easterly weather along the Queensland coast, reducing clarity and colour, and making it difficult to recognize anything with certainty. However, we thought we could make out the sandy hill behind Point Hicks, and hoped we might find a temporary anchorage in the lee of that low headland. But after we had changed down from the No. 2 to the No. 3 jib because of the freshening wind, we made poor progress, and the unrelenting current continued to set us to leeward, so that by the time we were nearing the shore Point Hicks was 15 miles to windward of us, and for the time being therefore out of reach. As in these conditions the wind is often at its strongest in the afternoon, we hove-to on the off-shore tack and waited for it to moderate.

But it did not moderate; instead it freshened to between 30 and 40 knots, and it seemed so absurd to lie there in discomfort slowly losing ground, that after supper of corned beef sandwiches and a drop of scotch, we decided to make use of that wind and go to Port Phillip Bay after all, and we felt glad that we did not have to explain this volte-face to anyone but ourselves. So we rolled a reef in the mainsail and squared away. Immediately the motion became more easy and the wind no longer shouted so loudly in the rigging, and we made good speed through the night, during which we passed two big clusters of lights, which we guessed (correctly, as we discovered later) were on oil rigs which were not shown on our old chart. The following afternoon, as the wind fell light, we approached Wilsons Promontory (the southernmost tip of the Australian mainland), and fortunately identified it before all except its highest part became for a time wreathed in fog. How good were the early navigators at giving names! For example, Skull Rock, one of the islands off that fine headland, does look just like a skull when viewed from the west, and it is unfortunate that on the modern chart its name has been changed to Cleft Island.

Ghosting, and occasionally motoring, through the night, which produced almost continuous lightning, some heavy rain, and a squall or two, we made better progress than anticipated, and it seemed that we might reach Port Phillip Heads before noon. I found the tide table puzzling, and could not decide exactly what 'Slack water flood, stream turns' meant. But Susan, using feminine logic, reckoned that if we made the Heads by 1100 we might get in before the ebb started to rush out, and that was important because the streams in the narrow entrance to the huge bay—almost an inland sea, for it is about 25 miles in diameter—can run at up to 8 knots, and because of an irregular bottom produces a tide race which can be dangerous. *Wanderer* just managed it, but I must say that in daylight the three lower leading-light structures, which indicate the centre and the sides of the entrance channel, were difficult for the stranger to identify, and from a distance even the high light, which is the back leading-mark, had competition from something that might have been a water tower.

We had been told that the anchorage in Camerons Bight off the southern shore is the best in all Port Phillip Bay, but that is not saying much because the bay does not offer a quiet, landlocked anchorage anywhere; that is the reason why few cruising yachts visit it or are stationed there, and the yachting accent is on racing. We made our way to Camerons Bight through a well-marked inshore channel, and were puzzled a bit by an odd-looking black buoy which on close approach proved to be the boiler of a wreck. Suddenly the gloom and the swell of Bass Strait were behind us; the sky cleared and the sun shone on golden beaches backed by gaily-painted houses; many sails leant to the breeze, and there were small and large craft everywhere with their people watching, for we had unknowingly arrived during one of the races for the Little America's Cup, an international event for catamarans. But, not being Melbourneites, the Bight was scarcely our idea of what a comfortable anchorage should be; the wind was from nor'-nor'-west, and the nearest land in that direction lay about 25 miles away and was invisible. No doubt the shoals round the Mud Islands did provide a little shelter from that direction, but we felt a lot happier when soon after nightfall there was a sudden weather change, and the wind fairly roared out of the south, coming then from the shore which was only a few hundred yards away.

In the calm of the morning we motored 30 miles north to Sandringham Yacht Club marina, and were given an outside pen which was a bit uncomfortable when the wind blew hard from the south, as it sometimes did during our stay, for the weather was unsettled and on one afternoon 3 inches of rain fell in 1 hour and 20 minutes to cause extensive flooding in the city. The club was kind, and two members took it upon themselves to bring us each morning bread, milk, and the day's paper. Many people showed great interest in *Wanderer*, and 35 visitors aboard in an afternoon was not unusual. Although the facilities were handy, we never managed to have a shower or do the laundry because we could not get along the walkway of the marina without meeting someone who wanted to talk or come on board. Such enthusiasm was, of course, not for us personally but for the way of life we represent, the dream of so many people less fortunate than ourselves. Towards the end of our highly social stay we were feeling a little grubby and jaded, so in another calm we motored down to Camerons Bight to wash ourselves and do the laundry, and to enjoy an undisturbed evening.

From the tide tables Susan had extracted the time to pass the Heads on the first of the outgoing stream in the morning, but as we approached that narrow channel thick fog rolled in from seaward. The high light, or something near it, remained visible long enough for us to get it on the correct bearing over the stern, and for a short time we could see a useful little island (which the chart shows as a drying rock) with a beacon on it. We steered a course to take us away slightly to the east of the main channel, so as to be clear of any shipping that might be taking advantage of the slack water, and it was as well that we did, for in quick succession the blurred, grey silhouettes of two large ships groping their way in, loomed up very close to starboard. Between the mournful blasts of the diaphone fog signal on Point Lonsdale we could hear the swish of the ships' bow-waves and the dull beat of their propellers.

How glad we were to get safely out and turn to the south-east, where we met other ships steaming in circles while they waited for the stream at the Heads to turn. There was no breath of wind, and we were not prepared to hang around and wait for one as it seemed more sensible to get clear of Bass Strait while we could and before the next gale arrived. So we motored on to round Wilsons Promontory by night, but as by then the fog had cleared, the nearly full moon was shining from a cloudless sky, and we knew what Skull Rock looked like, we had no difficulty in passing between the promontory and the row of unlighted islands south of it. At Melbourne we had bought a few more up-to-date charts, and from one of these we learnt that a traffic separation area (such as is used in the Straits of Dover) had been established off Wilson's Promontory. East-bound traffic, like ourselves, was supposed to pass south of the row of unlighted islands—Rodondo, Forty Foot Rock, etc—but by night and without the help of radar that could be difficult, and we noticed that other ships going east that night did the same as we did and ignored the one-way traffic arrangements.

The calm continued unbroken until the evening of the following day; then as we were passing between the oil-rigs *Halibut* and *Marlin*, the wind at last sprang up but unfortunately came from right ahead. We stopped the engine, and all through the night and the following forenoon we beat wetly on, making less and less progress against the increasing wind and sea, and at noon, when we were 15 miles to leeward of Point Hicks, we hove-to on the offshore tack with the north-east wind gusting to 35 knots. The conditions were just the same as we had experienced in the same position two weeks earlier, and we wondered if they were ever any different, or whether Gabo Island was a local Cape Horn.

For 46 hours we lay stopped, drifting through the water in a south-easterly direction, but probably being set by the current back into Bass Strait. Although the sun was shining, visibility grew poor, and we soon lost sight of the land. As we sipped the soup, which was all our squeamish stomachs could manage for supper, it was strange to recall that at that same time only four evenings before we had been giving a slide show to an audience of 700 in the comfort of the motionless club house at Sandringham.

When eventually the wind took off enough for us to resume beating to windward we were 80 miles south of Gabo Island, but we were 100 miles out in the Tasman Sea before we could effectively head north on the other tack. As we were then so far from the shore it seemed absurd to go in and visit Eden, for we were now near or on the sailing-ship route from south of Tasmania to Sydney, and in all probability outside the influence of the current which sets strongly south along the coast of New South Wales. So we stayed out there while we made our northing, and celestial observations showed that we had indeed and at last reached an area where there was no measurable current.

I have always found it difficult to judge with any degree of accuracy the strength of the wind, and I know I tend to exaggerate it, which is why we carry an anemometer; but the height of the sea is just as difficult to estimate and even easier to exaggerate. It was therefore of some interest to listen, as we made our way north in better weather, to the reports from lighthouses, etc, including those from the oil-rigs *Kingfisher A* and *Halibut*. Those rigs were only 14 miles apart, so one might have expected their reports of wind and sea conditions to be similar; but they were not. My log notes that one afternoon *Halibut* reported a north-east wind of 25 knots with a 6- to 8-foot sea, while *Kingfisher A*, which was within sight of *Halibut*, said the wind was east at 10 knots and the sea was 2 to 3 feet high. Perhaps *Kingfisher*'s observer had been mellowed by a good lunch and a bottle of plonk.

For all of two days, closehauled and mostly under all plain sail, our ship steered herself with the wheel lashed. It was purely by chance that we discovered her ability to do this, which was something I had not thought possible except with Slocum's remarkable old *Spray*, for I had always believed that the only way to achieve self-steering (here I am not considering wind-vane gears or auto-helmsmen) is to arrange a balance of sails and leave the helm free, so that a freshening or easing of the wind does not cause the vessel to luff or bear away. But clearly I was wrong so far as *Wanderer IV* was concerned, though I must admit that she will not always perform like that.

As we approached the latitude of Sydney we altered course inshore to cross the coastal current at a big angle so as to pass through it quickly, and while still more than 50 miles away we could see the loom of the city's lights in the overcast sky, but the first real light we saw was the brilliant flash of Macquarie when it was 20 miles off. By then the wind at last had shifted and was freshening from the south, and for the first time on that passage we eased sheets and flew before it. By dawn that wind had turned into a gale, and deciding that Broken Bay, 20 miles farther north, would suit our purpose better than Sydney in such conditions, we ran for it in the steep sea where wind and current were opposed, and at 0915, 7 days out of Port Phillip (the shortest distance by sea is 551 miles, but we had sailed 755), we rounded Barranjoey Head and came quickly into smooth water. Even there behind the land the wind was strong, but we remembered from our visit 18 years before that Refuge Bay offers secure anchorage in any weather, so we went on in, let go our anchor on firm mud, and while the wind blew and the rain fell we slept and slept. But as soon as the worst of the gale was over we made our way up Pittwater, one of the arms of the inlet, and our old friend Bob Godsall arranged for us to have a mooring next to his own yacht in Careel Bay.

When we had first met Bob on our earlier visit he had been the pilot for the Commonwealth Bank, but that job had terminated rather suddenly with the sale of the DC3 following a flight he had been ordered to make all the way from Sydney to Perth to bring back two nuns; this was a round journey of more than 4,000 miles. 'They sat erect like a couple of penguins,' Bob told us, 'and even the steward could not get a word out of them'. Now he owned a 4-passenger Beechcraft Bonanza with which he was doing charter work, and had recently carried Ann Todd on a film-making tour of Australia. He invited us and another couple for a flight.

'Now look,' said Susan over breakfast on the morning we were to join Bob, 'just relax and take it easy. No matter if something goes wrong, just remember you are not the navigator, you don't have to repair anything that busts. It's Bob's day, so let him do the worrying.'

It may sound unlikely, but Bob knew of an uninhabited island in a large lake about 200 miles north of Sydney, and that it had an abandoned air-strip which was reported to be still in good condition, and he flew us there for a picnic lunch of prawns and beer. But on the way there had been some trouble with the alternator which was not charging properly. After we had eaten we walked about for a short while, avoiding the hornets' nests, and then climbed aboard for the return flight. Bob pressed the starter switch and the propeller turned slowly, then slower still, and finally stopped.

'Battery's dead,' he said as he got out and tried, by swinging the propeller, to make the engine catch; but there seemed to be no future in that, so he got back into the pilot's seat and lit a cigarette. I noticed there were beads of sweat on his forehead, and I remembered Susan's words: 'It's Bob's day, let him do the worrying.'

'What happens next?' I asked.

'Well, normally one would just ring the nearest Beechcraft agent and have some joker bring a fresh battery out, but here . . .?' I saw his point, and lit my pipe.

A remarkable thing then happened. Out from behind the bush bordering the air-strip rumbled a bulldozer with a big man driving and a woman (his wife?) sitting in the stern holding a broom. They seemed as surprised to see us as we were to see them. It turned out that the island had recently changed hands, and the new owner intended to open it up as a tourist resort, and as the first thing was to tidy it a bit, the bulldozer had been landed from a barge to make a start. One might almost have supposed this to have been arranged specially by Providence for our benefit. In little careful jerks the yellow monster was backed to within a few inches of the fragile blue and white plane—beauty and the beast!—a jumper lead was connected from the bulldozer's battery to the plane (Plate 30, *bottom*), and within a few moments the propeller was spinning and the exhaust crackling healthily. We fastened our seat-belts for the second time and flew, following the line of long golden beaches separated one from another by bluff headlands, south past Port Stephens and Newcastle, towards the urban sprawl of Sydney, and on the way we did a quick circle over the bay where *Wanderer* lay quietly at her mooring.

During the passage round from Port Phillip our steering had developed a tight spot, and with Bob's help we traced this to a slightly bent shaft in the Mathway gear; we believed this must have been caused by the heavy strains imposed while we were lying hove-to. As it was impossible to do anything about it without tearing out the joinery work in the sleeping cabin, we agreed to leave it until we should return to New Zealand, but we did have the massive steel emergency tiller altered so that we could ship it on the squared rudder head without having to move the dinghy, and we decided that if we had to lie-to during the trip we would ship the tiller and lash it with springy nylon rope so as to take all the strain off the Mathway gear.

But first we had to sail down to have a look at Sydney and get our customs clearance, and when we stuck our bowsprit outside the shelter of Broken Bay we found an enormous swell running and breaking in a spectacular fashion along the coast. This was caused by a tropical disturbance, of which there had been a succession during our time in Australian waters, off the Queensland coast. There was a gloominess, a sinister air, about the scene, and very little wind, and we were glad to motor into the smooth water of Sydney Harbour, and after having a look at the familiar old bridge and the gleaming new opera house, which had cost a fortune and still was not complete, we went into Rushcutters Bay and got a berth in the marina of the Cruising Yacht Club. A friend drove us in to renew our acquaintance with the thriving, throbbing, cosmopolitan city, parts of which were in process of being torn up by earth-moving machines to make an underground railway, and again we showed slides and talked, topped up with water and provisions, and bought from the club two cartons of Australian beer, which we had always thought to be the best and the most potent in the world.

We slipped out of Sydney's great harbour at the end of March, by which time the cyclone season should be over, and so hazy was the day that the land disappeared astern when we were only 5 miles from it.

'Farewell, Australia!', wrote Charles Darwin aboard the *Beagle*, 'you are a rising child . . . but you are too great and ambitious for affection, yet not great enough for respect. I leave your shores without sorrow or regret.'

That was pretty damning, but I believe Darwin's thoughts on a similar occasion today would be different. Certainly ours were, and we were sad to be leaving behind that vibrant, virile country of contrasts and colours and great open spaces, and particularly our many friends there. But this is a sadness that all voyaging people have to endure again and again, though when the land has faded out of sight I believe that most of us tend to concentrate more on the passage and the problems that lie ahead rather than on regrets at parting; this, however, does not imply that we quickly, or indeed ever, forget the people who have been so kind and generous to us, and have implanted a feeling of warmth that is good to have around one, particularly when sailing conditions are difficult, and one is wet and cold, and possibly a little scared.

It is more than 1,000 miles from Sydney Heads to New Zealand's North Cape, and on the second day at sea the wind, which had been light and fair, shifted round and came ahead, and we were jammed hard on the wind all the rest of the way, and for much of the time could not even lay the course; progress was therefore slow and abominably uncomfortable. Going to windward in the open ocean aboard any yacht must be a miserable business, and certainly life in *Wanderer IV* was no exception; indeed, on this passage the jolting and the noise at times so exhausted us that on several occasions we had to stop to get some rest. There were some little irritations, too: one day I forgot to close the valve on the galley sink, and as we took a particularly violent plunge, a jet of greasy water shot up and hit the white coachroof ceiling. When I punctured the first can of our Australian beer, its seemingly explosive contents also sprayed the ceiling; we tried many other of our 96 cans, all were the same and none was drinkable, but fortunately we did have a bottle or two of wine. On one wild night the mizzen sheet got washed out through a scupper, and became almost inextricably tangled with the log-line.

Although, as I have said, the cyclone season should be finished by the end of March, we learnt from the radio when we tuned in to get the A.B.C. news, that a late arrival, Emily, was making a nuisance of herself farther north, where she had already sunk two of the yachts taking part in the Brisbane to Gladstone race with her 110-knot wind. After that we listened in frequently to learn what she was doing, for often these disturbances recurve on nearing the Queensland coast and make their way to the south-east to cross the course that we were following; we were immensely, though selfishly, relieved when we heard that she had gone inland and turned into a rain depression, and was flooding large areas of the countryside. Nevertheless Emily did eventually recurve to have a go at us, but as she had by then used up most of her energy she only brought along a wind of 40 knots, but that was more than enough to persuade us to heave-to. So, as agreed earlier, we shipped the tiller and lashed it so as to take all strain off the steering gear. There was a lot of water sluicing over the stern deck that black, wild night, and this was added to by the deluge of rain. Some gallons of this naturally found its way into the afterpeak through the deck hole which normally was covered by a screw-in plate, where the tiller was shipped on the rudder head, and after the storm had left us we spent many hours getting rid of the oily water and drying the gear that was stowed there. We had hoped that the departure of these tattered remnants of a cyclone would bring a change of wind, but it did not, and the wind remained persistently ahead, and we therefore started once more to crash to windward.

After Emily had gone we had a passenger. A noddy alighted on the stowed mizzen, stayed there for the night, and returned the following evening. We do not like sea-birds to come aboard because it usually means there is something wrong with them, and through lack of the right food, or seasickness, they grow weak, and get drowned in a miscalculated attempt to take off or alight, and of course when Nicholson was with us he used to kill or damage them. But this bird seemed to be in good fettle, and spent much time preening itself. On the third night there were two of them roosting on the mizzen and twice the quantity of droppings to clean up. Our attempts to shoo them off were not successful, for they were completely unafraid, so I put my hand under the tail of one and gave it a push; it was an odd sensation, for the bird did not flutter or use its legs to get air-borne, it simply spread its long, brown wings and soared away apparently right into the wind's eye—I wished we could go to windward like that. It was back within a few minutes, preening alongside its friend, and I let it be; but in the morning we set the mizzen, and that discouraged our feathered boarders, though they did make a few half-hearted attempts to land elsewhere.

In the forenoon of our eleventh day out from Sydney we sighted the Three Kings, a group of uninhabited islands which lies about 30 miles north-west of Cape Reinga. There was the usual strong headwind and steep sea, and on the starboard tack we stood away to the north-east. We tacked at midnight and hove-to until dawn, then continued sailing, and when we again had the Three Kings within sight, we found that in 24 hours we had made good a distance of only 10 miles over the ground. I began to understand more clearly how a sailing vessel could become embayed; that had happened to the Canadian yacht *Endeavour* on the New Zealand coast only a short time before.

Heavy rain then set in, but we managed to find North Cape light during the night, and having passed it we at last got a change of wind, and for the first time in nine days could ease the sheets a little and sail at a respectable speed along the familiar coast, past Mount Camel, Whangaroa, and the Cavallis; then another downpour obscured everything, but the Ninepin appeared when and where it should, and in the evening, escorted by a joyous party of porpoises, we entered the Bay of Islands as the sun came out. As we let go the anchor off the bright little town of Russell, which was circled by a rainbow, a clear hail came from the shore:

'Welcome home, *Wanderer IV*.'

# Appendix: Details of Wanderer IV

The yacht was designed by Mr. S. M. Van der Meer, built by Mr. J. Jongert at Medemblik in Holland, and launched in April 1968. Her dimensions are: l.o.a. 49·5 ft; l.w.l. 40 ft.; beam (not including rubbing strakes) 12·5 ft.; designed draught 5·75 ft., actual draught 6·25 ft. (but see below); displacement 20 tons.

The hull is of heavy-displacement type with about 20 feet straight of keel, a clipper bow, and a 'sawn off' counter stern. Ballast was about 4 tons of lead, which I felt sure was not enough as the yacht was more tender than one of her size and type should be. However, since the voyages described in this book were made an additional 2 tons of ballast has been added on the under side of the keel. This has proved to be a great improvement, and the yacht now stands up to her canvas well, but the alteration has increased her draught, to 6·75 ft.

The clipper bow has proved to be buoyant, though not too much so, and checks the pitching gently; but unfortunately the bowsprit casing is square in section instead of being round, and over its inboard end has a solid platform to enable the jibs to be handled with ease; and each side of this platform are rectangular steel boxes carrying the sheaves for the anchor cables. When the yacht pitches all that lot into the sea, as she did many times a minute during the Tasman crossing, the flat surfaces hit the water with such force that the blow can be felt throughout her, and tends to stop her forward progress. I do not know why most of the steel yachts with which I am familiar, including *Wanderer IV*, have wide rubbing strakes at deck level throughout most of their length, for wood and g.r.p. yachts do not appear to need these. When lying alongside another vessel or a dock this protrusion makes it difficult to keep fenders in place, and if fenders are not used the paint will get damaged and rust will form unless the strakes are shod with stainless steel. When sailing with the wind abeam or forward of the beam the lee strake slams heavily. It seems to me that a sailing vessel should be so arranged that she presents the minimum of flat surfaces to the sea against which she will so often be trying to make progress, unless she is doing a circumnavigation in the trade winds.

The hull, main deck, and coachroof coamings are of steel, all welded, the plating being $\frac{5}{32}$ in. thick, and the $1\frac{3}{4}$-in. frames spaced 15 in. apart. The main deck has plywood bolted to it, and on top of this is fastened a deck of laid teak with the planks following the curve of the side, and this is carried right out to the bulwarks so that there are no waterways. This looks attractive and is pleasant to walk on barefoot, but the total weight of the deck must be considerable and was probably a contributory factor in the yacht's original tendency to be crank. The coachroof decks are of g.r.p.-covered plywood on steel beams. In way of the living spaces the under sides of the main deck and coachroof decks, and the topsides down to the level of the cabin soles, are all insulated and ceiled with plywood; this keeps the accommodation cool in hot weather, prevents condensation, and reduces noise; except in the engine-room there is no visible sign below that this is a steel vessel. All hatches, rails, doors, etc, are of teak, or something that looks rather like it; joinery work below is of plywood with teak veneer, some of which we have painted white. The forehatch, and the skylights over the galley and sleeping cabin, are of the non-leaking Maurice Griffiths type with double coamings; all have side-flaps, and all can be hinged fore or aft so as to extract or blow. There are 13 opening portlights in the coachroof coamings, each measuring 12 by 6 in., some cowl vents, and a windsail for use in the tropics. All openings are provided with easily removable mosquito screens.

The forepeak, with a hatch in the coachroof deck above it and a door from the saloon, can be used as a workshop or a photographic darkroom, and has an electric lead from a dynamotor in the engine-room to bring through a 240-volt supply. It

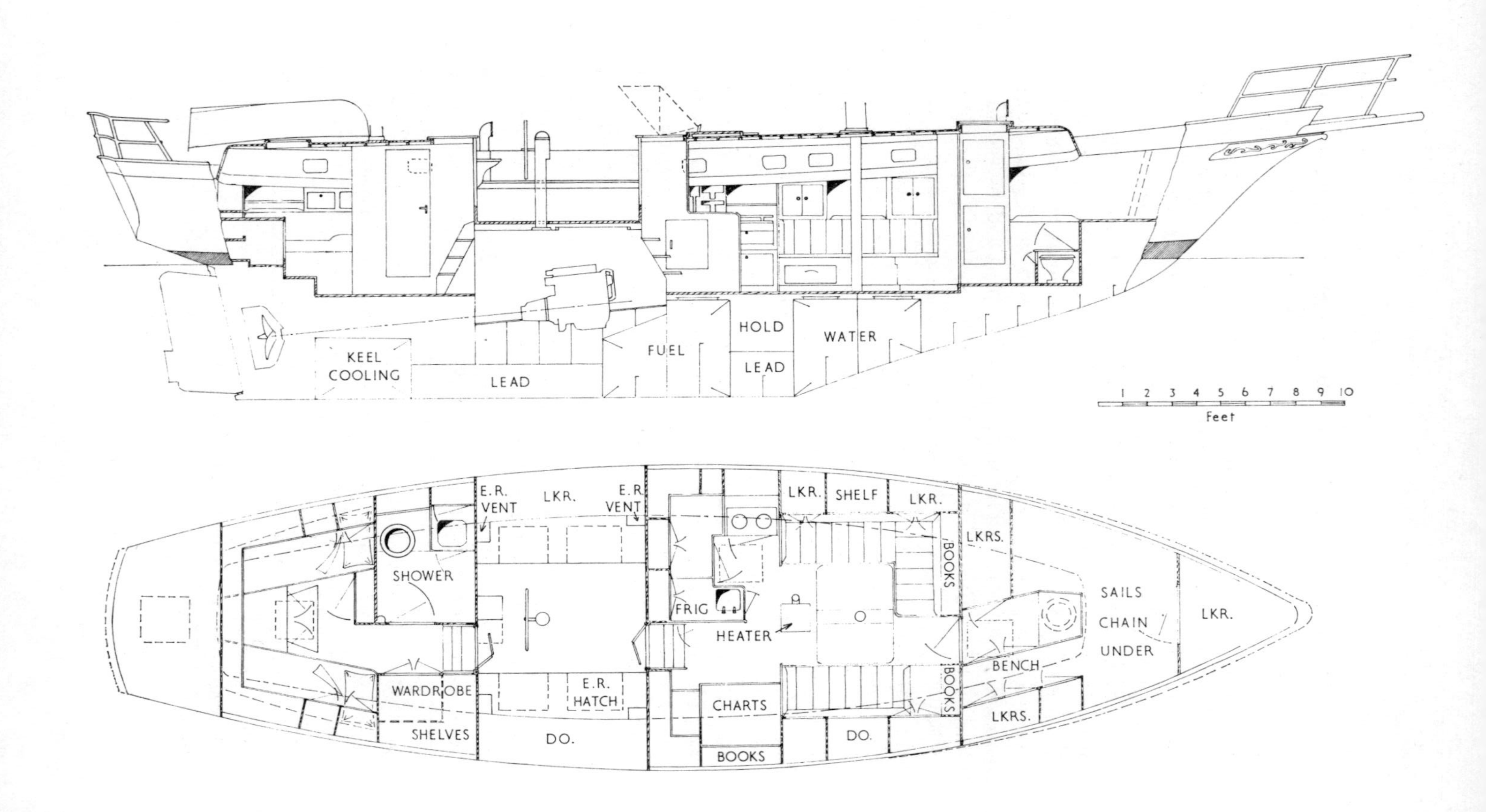

HOLD
WATER
KEEL
COOLING
LEAD
FUEL
LEAD
1 2 3 4 5 6 7 8 9 10
Feet
E.R.
VENT
LKR.
E.R
VENT
LKR.
SHELF
LKR.
LKRS.
BOOKS
SHOWER
SAILS
CHAIN
UNDER
LKR.
FRIG
HEATER
BENCH
BOOKS
WARDROBE
E.R.
HATCH
CHARTS
LKRS.
SHELVES
DO.
DO.
BOOKS

contains the chain locker and ample stowage for all sails, tools, spares, and photographic equipment. A Baby Blake heads is installed there.

The saloon has a central table round the mainmast support; to starboard is a settee with a sideboard at its forward end, and to port is an L-shaped settee. The forward bulkhead is fitted with bookshelves, and there are lockers, drawers, etc, each side above the settee backs, both of which have stowage outboard of them. A Kempsafe diesel heater is fitted abaft the table, its header tank being in the engine-room entrance. At the after end of the saloon on the port side is the galley, with its sink on the centre line, and a swinging, 2-burner Taylor's Para-Fin cooker, on which a small oven can be fitted; there are many lockers, stowage for crockery, drawers for cutlery, a large, Formica-covered working area, and an Electrolux, absorption-type refrigerator, which burns less than a gallon of paraffin a week, and has its flue vented to the deck. At the aft end of the saloon on the starboard side is an oilskin locker, a chart table with a bookshelf outboard of it, and stowage beneath it for at least 800 charts. Double-bottom tanks, two for water and one for diesel fuel, are under the saloon sole, the hold between them being used for the stowage of cases of canned food. The water tanks hold about 100 gallons each; the capacity of the fuel tank is unknown, but is probably in excess of 200 gallons. Fillers for the tanks are in the side-decks, and the air-vent pipes emerge in the engine-room ventilation trunks.

Three steps lead up to the central self-draining cockpit, which is long enough to sleep in, and under its fore-and-aft seats has lockers extending out to the topsides. The steering-wheel pedestal, on which is the compass and the controls for the engine and the auto-helmsman's clutch, is towards the aft end, where there is a folding seat and footrest for the helmsman. The forward part of the cockpit and the saloon companionway are protected by a folding hood. The engine-room, with nearly 4 feet of headroom, is situated below the cockpit deck; it is soundproofed, and is separated from the living spaces by steel bulkheads; access to it is by way of one of the lifting cockpit seats, and there is no passage through from saloon to sleeping cabin. In the steel deck under the removable wooden cockpit deck is a large, bolted-down hatch, which would enable the engine to be lifted out if that should ever be necessary.

The engine is a 4-cylinder Ford diesel, which has a continuous rating of 61 h.p. at 2,400 r.p.m.; it drives a 20-in., 2-blade propeller through a Borg Warner hydraulic gearbox with 2 to 1 reduction gear, and a brake, operated from on deck, locks the shaft when sailing. We have not run at full speed except over a short, measured distance, when we got a speed of 7½ knots. Normally we run either at 1,500 r.p.m., which gives a speed of 5 knots in smooth water, or 1,800 r.p.m., which gives a speed of 6¼ knots. At the lower speed the fuel consumption is about 0·85 gallon per hour. If we take the usually accepted figure for diesel engines of 0·4 pint of fuel per horse-power/hour, it would seem that the engine develops about 18 h.p. when run like that. Lead-lined steel boxes each side of the engine-room hold the lead-acid batteries, which provide a voltage of 24 and a capacity of 400 ampere/hours. The Ford has a 25-amp alternator, but there is an auxiliary charging plant, a single-cylinder Yanmar diesel engine driving a 50-amp generator. Both engines are flexibly mounted and freshwater-cooled, the keel cooling-tank being right aft, and the water has soluble oil mixed with it to deter rusting. Aft on the port side of the engine-room is a Sihi water-pressure system which is worked automatically by a pressure-controlled electric pump. Near it is the auto-helmsman's electric motor, which drives the Mathway steering gear through a chain and sprockets. Suction pipes from each of the four bilges come to a manifold in the engine-room, through which any bilge can be pumped by a hand-operated Whale diaphragm pump, or by a Jabsco impeller pump belt-driven by the Ford; this same pump can be used to deliver salt water on deck for scrubbing down or washing the cable.

From the after end of the cockpit steps lead down to the sleeping cabin, which has a bunk each side, lockers, shelves, and drawers for clothes and books, and a large hanging wardrobe and linen locker. In a compartment to port is the heads, another Baby Blake, but this is the de luxe model which is a bit more generous than the utility model forward. There is a wash-basin, a shower head (of the 'telephone' type) with gas heater, the gas cylinder being in a steel locker, which drains overboard, under a cockpit seat; a stainless tray under teak gratings collects shower water, and is emptied by a small electric Jabsco pump. Above the wardrobe and handy to the helmsman are the switches for the navigation lights together with the control box for the auto-helmsman; the magnetic heading unit for this is under the cabin steps.

The afterpeak, reached by a flush-fitting hatch in the deck, is separated from the cabin by a steel bulkhead; the rudder stock and gland, over which I have sweated for so many hours, together with part of the Mathway linkage, is situated there, and is protected by portable wooden covers, so that fenders, lines, warps, boarding ladder, etc, stowed in the afterpeak, cannot become entangled with the steering gear.

The ground tackle comprises a 75-lb CQR (plough-type) bower anchor on a 45-fathom length of Lloyds-tested ½-inch short-link chain with a breaking strain of 6 tons; the kedge is a 45-lb CQR on a 2¼-in. circumference nylon warp; and a 100-lb fisherman-type anchor (originally made for the R.N.L.I.) is carried on the fore-deck as a stand-by or hurricane anchor. The bower chain is handled by a Simpson Lawrence electric windlass, which also has a barrel for handling rope, and in the event of an electrical failure it can be worked powerfully, but very slowly, with a hand lever. At the stemhead 7-in. diameter nylon sheaves accommodate chain and warp, but nylon is not the ideal material for such a purpose because in time it tends to spread and bind under load. The bower does not have to be lifted inboard, as it stows with its shank lying on a small raised deck just abaft the stemhead, and with its plough wedged under the bowsprit, which is protected by a steel plate. For making fast when lying alongside, and for other purposes, there are eight 3½-in. diameter stainless bollards, each with a pin through it, welded to the steel deck—one on each bow, one on each quarter, and a pair each side amidships; the edges of the fairleads in the bulwarks which serve these, and on which no paint will stand, are shod with stainless steel. All stanchions, guardrails, etc, are stainless, and the male fittings for the stanchions are welded to the steel deck. Each side of the cockpit a section of the guardrails can be opened by means of senhouse slips, but aft of these, and right round the stern, the upper guardrail is not of wire, but is a flat stainless bar covered with teak; this may not be very practical, but it does look and feel pleasant, and we have not damaged it yet.

*Wanderer*'s rig, which was not of our choice, is that of a ketch, but with a split fore-triangle, which was our choice. Both the masts, which stand on the specially strengthened coachroofs, and the mizzen boom, are of spruce; they are hollow and of rectangular section; the main boom is also of spruce, but is solid and round in section, and is fitted with Lewmar worm-type roller reefing gear. The bowsprit is of steel, sheathed in teak, and is provided with a chain bobstay. With that single exception, all the standing rigging is of 1 × 19 stainless wire with swaged terminal fittings; all mast and deck fittings are also of stainless steel. Originally all rigging screws were stainless with closed barrels, but 12 of the larger ones serving the mainmast proved unsatisfactory and were replaced with the open-barrel type made of bronze. The halyards are of Marlow pre-stretched 1½-in. 3-strand terylene, and lead to Gibb open-barrel winches; sheets also are of terylene, most of them being of the plaited type. There is a pair of Gibb 9CR two-speed winches for handling the jib-sheets, and a similar winch for the mainsheet. The mizzen sheet is rigged as a 3 to 1 purchase, which often is not powerful enough, and a 4 to 1 tackle is clapped on the staysail sheets when needed. Both topping lifts are of nylon. Two shifting backstays, or runners, for the mainmast when the staysail is set, are handled by Lewmar levers; mizzen runners, required only when the mizzen staysail is set, are set up by tackles.

All the sails were made by Cranfield of Burnham-on-Crouch, and the working sails are of tan-dyed terylene; light weather and running sails are of white terylene. There are three jibs: No.1 (or genoa) for light winds; No.2 (or yankee) the normal working jib; and No.3 (storm jib) for heavy weather. With ketch rig the staysail, which is one of the most efficient and easily handled of all sails, is unfortunately the smallest of the working sails; originally it had a boom, but that was scrapped early on the voyage. Mainsail and mizzen both have straight leeches (no roach), and had no battens; but a new mainsail that we had sent out to New Zealand does have battens, and they have not caused any serious problems so far. The white plastic slides with which the sails originally were fitted proved unreliable, possibly due to an excess of sunlight, and many of them fractured; slides of black plastic seem to have a longer life. Neither sail has slides on the lower third of its luff, rope lacings being used there to make reefing and unreefing quicker and simpler. Originally we had a pair of running sails, the alloy booms for which were pivoted at the root of the lower crosstrees. But, as has been mentioned elsewhere in this book, we found that rig unsatisfactory; now we use only one of the sails, and its boom is pivoted low on the mast.

Closehauled in a moderate wind we can carry 1,040 square feet of sail, but on a reach can increase this to 1,640 square feet. The sail areas, and weights of cloth (per square yard) are:

| | | |
|---|---|---|
| Mainsail | 350 sq.ft. | 10 ounce |
| Mizzen | 208 sq.ft. | 11½ ounce |
| Staysail | 182 sq.ft. | 10 ounce |
| No.1 jib | 500 sq.ft. | 6 ounce |
| No.2 jib | 300 sq.ft. | 8¼ ounce |
| No.3 jib | 135 sq.ft. | 11½ ounce |
| Mizzen staysail | 400 sq.ft. | 3 ounce |
| Running sail | 270 sq.ft. | 6 ounce |

It will be noticed that the mainsail is of lighter cloth than the mizzen, but this is because it is a newer sail and the original weight of 11½ ounces was unnecessarily heavy. When the time comes for a new mizzen, that too will be of lighter cloth, for I am convinced that heavy sails are a mistake.

Commensurate with the size of our vessel and her small crew, we tried to keep things as simple as possible, and perhaps this is reflected in the list of instruments that we carry. These are: A Sestrel grid-type steering compass with luminized grid; a simple, home-made sight ring stands on top of the bowl when bearings have to be taken. A Brookes & Gatehouse echo-sounder is mounted in the cockpit, and a Zenith radio receiver in the saloon; we carry no transmitter. We do have a Brookes & Gatehouse D/F set, but as mentioned on page 15, this cannot be used properly in a steel vessel, where one needs the type with a loop mounted in a fixed position so that it may be calibrated. We have none of the usual electronic gear for telling the strength and direction of the wind, or the speed through the water and the distance run, but we do have a hand-held anemometer, which we believe is very conservative, and an old and trustworthy Walker Excelsior patent log. We have an equally old aneroid, and a much older barograph. The sextant is a Husun with micrometer head, and the chronometer an ancient Waltham watch slung in gimbals.

A g.r.p. 7½-ft. dinghy, built by Souter of Cowes, is carried capsized over the cabin skylight, and is hoisted in or out by the mizzen halyard. A Seagull outboard motor stows, when not in use, in the engine-room.

In the four years that have slipped past since *Wanderer IV* was launched, she has made good 25,600 sea miles, in conditions ranging from calm to storm, from glass-smooth water to 20-ft., heavily-crested seas. We have done many things to her and for her, most of which turned out to be considerable improvements. As a voyaging home and workshop she serves us very well; on a rainy day in port she gives us room in which to run around and dry our clothes, and she provides us with a reasonably steady, airy space in which to entertain our friends; also, and Susan and I think this is important, she looks good in any nautical company. Naturally there are some things we wish were different or had never happened; but every ship is a compromise, most have minds of their own, and none is perfect, though there are very few owners who will admit this.

# Index